Human Factors Issues in Motorcoach Emergency Egress

INTERIM REPORT 1 - FINAL

PREPARED FOR:

Human Factors Engineering Integration Division, NVS-331
Vehicle Safety Research Office
National Highway Traffic Safety Administration
U.S. Department of Transportation
Washington, DC 20590

PREPARED BY:

John A. Volpe National Transportation Systems Center
Research and Innovative Technology Administration
U.S. Department of Transportation
Cambridge, MA 02142

August 2009

U.S. Department of Transportation

National Highway Traffic Safety Administration

Memorandum

Subject: Motorcoach Safety Research

Date: **DEC 0 2 2009**

From: Lori K. Summers
Director, Office of Crashworthiness Standards

Reply to Attn of:

To: Docket No. NHTSA-2007-28793

Thru: O. Kevin Vincent
Chief Counsel

Dorothy R. Nakama Nov. 18, 2009

Attached is a report titled, "Human Factors Issues in Motorcoach Emergency Egress - Interim Report," prepared by the John A. Volpe National Transportation Systems Center. This interim report presents the results of the first year of a two-year NHTSA-funded research effort to address human factors issues related to the emergency evacuation of motorcoaches and other human factors elements of Federal Motor Vehicle Safety Standards (FMVSS) No. 217 that apply to buses other than school buses.

This interim report includes a literature review that identifies regulations/standards and research studies from other transportation modes and other countries that may be applied to emergency egress-related requirements for buses. The report also presents preliminary findings on three emergency egress topics: 1) emergency exits, 2) interior and exterior emergency exit marking, and 3) emergency exit lighting. Potential motorcoach design changes identified and discussed in this interim report include: redesign of the wheelchair access door and /or installation of another side door for use as an emergency exit; redesign of emergency window exit opening and release systems; additional and larger emergency roof exit hatches; and increased conspicuity of emergency exits.

Please submit it to the docket NHTSA-2007-28793.

Attachment:

Human Factors Issues in Motorcoach Emergency Egress - Interim Report (Year 1)

Notice

This document is disseminated under the sponsorship of the Department of Transportation in the interest of information exchange. The United States Government assumes no liability for its contents or use thereof.

Notice

The United States Government does not endorse products or manufacturers. Trade or manufacturers' names appear herein solely because they are considered essential to the objective of this report.

REPORT DOCUMENTATION PAGE		*Form Approved* *OMB No. 0704-0188*

Public reporting burden for this collection of information is estimated to average 1 hour per response, including the time for reviewing instructions, searching existing data sources, gathering and maintaining the data needed, and completing and reviewing the collection of information. Send comments regarding this burden estimate or any other aspect of this collection of information, including suggestions for reducing this burden, to Washington Headquarters Services, Directorate for Information Operations and Reports, 1215 Jefferson Davis Highway, Suite 1204, Arlington, VA 22202-4302, and to the Office of Management and Budget, Paperwork Reduction Project (0704-0188), Washington, DC 20503.

1. AGENCY USE ONLY (Leave blank)	2. REPORT DATE August 2009	3. REPORT TYPE AND DATES COVERED Interim Report September 2007 – November 2008
4. TITLE AND SUBTITLE Human Factors Issues in Motorcoach Emergency Egress		5. FUNDING NUMBERS HS15/HS15A1/HS57A1
6. AUTHOR(S) John K. Pollard and Stephanie H. Markos		
7. PERFORMING ORGANIZATION NAME(S) AND ADDRESS(ES) Research and Innovative Technology Administration John A. Volpe National Traffic Systems Center U.S. Department of Transportation Cambridge, MA 02142-1093		8. PERFORMING ORGANIZATION REPORT NUMBER
9. SPONSORING/MONITORING AGENCY NAME(S) AND ADDRESS(ES) U.S. Department of Transportation National Highway Traffic Safety Administration Human Factors Engineering Integration Division Vehicle Safety Research Office, NVS-331 Washington, DC 20590		10. SPONSORING/MONITORING AGENCY REPORT NUMBER

11. SUPPLEMENTARY NOTES

12a. DISTRIBUTION/AVAILABILITY STATEMENT This document is available on the NHTSA web site at www.nhtsa.dot.gov.	12b. DISTRIBUTION CODE

13. ABSTRACT (Maximum 200 words)

FMVSS 217, *Bus Emergency Exits and Window Retention and Release* specifies a series of dimensional and physical requirements for emergency exits. The intent of NHTSA is "to minimize the likelihood of occupants being ejected from the bus and to provide a means of readily accessible emergency egress" for those occupants under crash and other emergency scenarios. These scenarios can include catastrophic bus accident situations, such as a vehicle fire, rollover, or water immersion where immediate emergency evacuation is necessary under life-threatening and difficult conditions.

In 2007, NHTSA issued a research plan to address priority actions specifically related to motorcoach emergency egress.

This interim report describes the preliminary findings of the first year of a two-year NHTSA-funded study which focused on three topics: 1) emergency exits, 2) interior and exterior emergency exit marking, and 3) emergency exit lighting.

A literature search was completed; several field visits were conducted, which included emergency window and roof exit hatch operation; and two sets of motorcoach egress experiments were conducted.

Potential motorcoach design changes identified and discussed in this interim report include: redesign of the wheelchair access door and / or installation of another side door for use as an emergency exit; redesign of emergency window exit opening and release systems; additional and larger emergency roof exit hatches; and increased conspicuity of emergency exits, either by use of high performance photoluminescent marking material or crashworthy emergency exit lighting, or dual-mode systems, which combine both technologies.

14. SUBJECT TERMS bus safety, motorcoaches, buses, school buses, NHTSA, FMVSS 217, emergency egress, emergency evacuation, emergency exits, photoluminescent, HPPL, emergency lighting			15. NUMBER OF PAGES 200
			16. PRICE CODE
17. SECURITY CLASSIFICATION OF REPORT Unclassified	18. SECURITY CLASSIFICATION OF THIS PAGE Unclassified	19. SECURITY CLASSIFICATION OF ABSTRACT Unclassified	20. LIMITATION OF ABSTRACT Unlimited

NSN 7540-01-280-5500

Standard Form 298 (Rev. 2-89)
Prescribed by ANSI Std. 239-18
298-102

METRIC/ENGLISH CONVERSION FACTORS

ENGLISH TO METRIC

LENGTH (APPROXIMATE)
- 1 inch (in) = 2.5 centimeters (cm)
- 1 foot (ft) = 30 centimeters (cm)
- 1 yard (yd) = 0.9 meter (m)
- 1 mile (mi) = 1.6 kilometers (km)

AREA (APPROXIMATE)
- 1 square inch (sq in, in^2) = 6.5 square centimeters (cm^2)
- 1 square foot (sq ft, ft^2) = 0.09 square meter (m^2)
- 1 square yard (sq yd, yd^2) = 0.8 square meter (m^2)
- 1 square mile (sq mi, mi^2) = 2.6 square kilometers (km^2)
- 1 acre = 0.4 hectare (he) = 4,000 square meters (m^2)

MASS - WEIGHT (APPROXIMATE)
- 1 ounce (oz) = 28 grams (gm)
- 1 pound (lb) = 0.45 kilogram (kg)
- 1 short ton = 2,000 pounds (lb) = 0.9 tonne (t)

VOLUME (APPROXIMATE)
- 1 teaspoon (tsp) = 5 milliliters (ml)
- 1 tablespoon (tbsp) = 15 milliliters (ml)
- 1 fluid ounce (fl oz) = 30 milliliters (ml)
- 1 cup (c) = 0.24 liter (l)
- 1 pint (pt) = 0.47 liter (l)
- 1 quart (qt) = 0.96 liter (l)
- 1 gallon (gal) = 3.8 liters (l)
- 1 cubic foot (cu ft, ft^3) = 0.03 cubic meter (m^3)
- 1 cubic yard (cu yd, yd^3) = 0.76 cubic meter (m^3)

TEMPERATURE (EXACT)
$[(x-32)(5/9)]\ °F = y\ °C$

METRIC TO ENGLISH

LENGTH (APPROXIMATE)
- 1 millimeter (mm) = 0.04 inch (in)
- 1 centimeter (cm) = 0.4 inch (in)
- 1 meter (m) = 3.3 feet (ft)
- 1 meter (m) = 1.1 yards (yd)
- 1 kilometer (km) = 0.6 mile (mi)

AREA (APPROXIMATE)
- 1 square centimeter (cm^2) = 0.16 square inch (sq in, in^2)
- 1 square meter (m^2) = 1.2 square yards (sq yd, yd^2)
- 1 square kilometer (km^2) = 0.4 square mile (sq mi, mi^2)
- 10,000 square meters (m^2) = 1 hectare (ha) = 2.5 acres

MASS - WEIGHT (APPROXIMATE)
- 1 gram (gm) = 0.036 ounce (oz)
- 1 kilogram (kg) = 2.2 pounds (lb)
- 1 tonne (t) = 1,000 kilograms (kg)
- = 1.1 short tons

VOLUME (APPROXIMATE)
- 1 milliliter (ml) = 0.03 fluid ounce (fl oz)
- 1 liter (l) = 2.1 pints (pt)
- 1 liter (l) = 1.06 quarts (qt)
- 1 liter (l) = 0.26 gallon (gal)
- 1 cubic meter (m^3) = 36 cubic feet (cu ft, ft^3)
- 1 cubic meter (m^3) = 1.3 cubic yards (cu yd, yd^3)

TEMPERATURE (EXACT)
$[(9/5)\ y + 32]\ °C = x\ °F$

QUICK INCH - CENTIMETER LENGTH CONVERSION

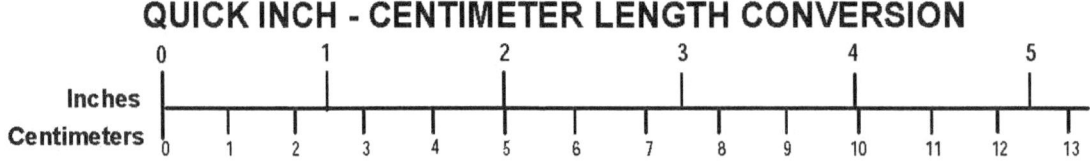

QUICK FAHRENHEIT - CELSIUS TEMPERATURE CONVERSION

°F	-40°	-22°	-4°	14°	32°	50°	68°	86°	104°	122°	140°	158°	176°	194°	212°
°C	-40°	-30°	-20°	-10°	0°	10°	20°	30°	40°	50°	60°	70°	80°	90°	100°

For more exact and or other conversion factors, see NIST Miscellaneous Publication 286, Units of Weights and Measures. Price $2.50 SD Catalog No. C13 10286

Updated 6/17/98

PREFACE

The mission of the National Highway Traffic Safety Administration (NHTSA) is to reduce motor vehicle crashes and injuries. NHTSA safety regulations for bus and school bus design are contained in Title 49, Code of Federal Regulations (49 CFR), Part 571, Federal Motor Vehicle Safety Standards (FMVSS).

FMVSS 217, *Bus Emergency Exits and Window Retention and Release* specifies a series of dimensional and physical requirements for emergency exits, including their size, location, opening forces, and marking; in addition to a series of release and retention tests for all windows, other than windshields. The intent of NHTSA is "to minimize the likelihood of occupants being ejected from the bus and to provide a means of readily accessible emergency egress" for those occupants under crash and other emergency scenarios. These scenarios can include catastrophic bus situations, such as a vehicle fire, rollover, or water immersion, where immediate emergency egress is necessary under life-threatening and difficult conditions.

In 2007, NHTSA prepared a comprehensive research plan to address motorcoach safety issues that identified several improvements for motorcoach design as priority items for consideration in future rulemaking. One consideration identified in this plan is to address items on the National Transportation Safety Board (NTSB) "Most Wanted List" of safety improvements, i.e., "easy opening windows and roof hatches that stay open during evacuations" (H-99-9).

NHTSA asked the John A. Volpe National Transportation Systems Center (Volpe Center) to provide human factors research, evaluation, and technical support to NHTSA to identify potential motorcoach design changes that may improve emergency egress.

This interim report describes the findings and topics for NHTSA consideration, as developed during the first year of the Volpe Center two-year study. A report containing final results and topics for NHTSA consideration will be completed after the Volpe Center performs the second year activities in 2009. While this effort is directed at intercity and charter / tour-over-the-road motorcoaches, any insights and information considered to be relevant to emergency egress requirements for any other large buses or school buses are also documented.

The three major topic areas addressed in this interim report are: 1) emergency exits, 2) interior and exterior emergency exit signage, and 3) emergency exit lighting.

Potential motorcoach design changes identified and discussed in this interim report include: redesign of the wheelchair access door and / or installation of another side door for use as an emergency exit; redesign of emergency window exit opening and release systems; additional and larger emergency roof exit hatches; and increased conspicuity of emergency exits, either by use of high performance photoluminescent marking material, crashworthy emergency exit lighting, or dual-mode systems, which combine both technologies.

ACKNOWLEDGMENTS

The authors of this report are *John K. Pollard* and *Stephanie H. Markos* of the John A. Volpe National Transportation Systems Center (Volpe Center), Research and Innovative Technology Administration, (RITA), United States Department of Transportation (USDOT).

The authors acknowledge the technical contribution and support of many individuals. Special appreciation is due to National Highway Traffic Safety Administration / US DOT, including *John Hinch*, Director, Office of Human-Vehicle Performance Research; *Michael Perel*, Chief, and *Stephen Beretzky*, Project Manager, Human Factors/Engineering Integration Division; as well as *Roger Saul*, Director, Office of Crashworthiness Standards; and *Charles Hott, David Sutula*, and *Lawrence Valvo*, of the Office Vehicle Safety Compliance, for their support and technical guidance and assistance, during the first year of the Volpe Center two-year study.

James Hansen, Michael Janovicz, and several other staff members at MGA Research Corporation, provided expert technical support related to egress in rollover crashes. *Peter Pan Bus Lines, Ritchie Bus Lines,* and *Wilson Bus Lines* provided access to motorcoaches in their fleets for photography and for measurement of release and opening forces.

Motor Coach Industries (MCI), Prevost Car Inc., and *Van Hool NV* supplied technical data for their motorcoaches.

Christopher Crean and *Thomas Lynch* of *Peter Pan Bus Lines*; *Michael Ritchie* of *Ritchie Bus Lines*; and *Michael Wilson* of *Wilson Bus Lines*; provided access to their motorcoach fleets for various Volpe staff field measurements, as well as many insights into emergency exit design and passenger emergency egress issues from the bus company operators' perspective.

The authors also acknowledge the following Volpe Center / RITA / USDOT staff who made valuable contributions to this study: *Stephen Popkin*, Chief, Human Factors Division; *Mary Stearns*, NHTSA Human Factors Program Manager; and the sixty Volpe Center staff members who participated as subjects during the motorcoach egress experiments conducted on June 12, 2008.

Michael Hughes, Director of Transportation; and *Peter A. Crossan*, Manager of Fleet Maintenance and Contract Compliance; *Boston Public School Department*; and *Cheryl Hinton*, Director of Bus Operations; and *James Monahan*, Deputy Division Chief, Transportation; *Massachusetts Bay Transportation Authority*, provided access to photograph school bus and transit bus exits.

In addition, *Marilyn Gross*, Volpe Center Reference Center, *MacroSys Research and Technology, Inc.*, provided important assistance with conducting the literature search; and *Richard Gopen, Multimedia Services / MicroLan Systems, Inc.;* provided critical video support and graphics assistance during the Volpe-conducted bus egress experiments.

Finally, the authors extend special appreciation for their major contributions during the report preparation process: *Cassandra L. Oxley, MacroSys Research and Technology, Inc.*, for editorial coordination; and to *Raquel Rodriguez,* Volpe Center; and *Barbara Siccone*, formerly of *Chenega Advanced Solutions and Engineering (CASE),* who both provided critical typing and formatting support.

EXECUTIVE SUMMARY

BACKGROUND

In 2007, the National Highway Traffic Safety Administration (NHTSA) prepared a comprehensive research plan to address motorcoach safety issues that identified several improvements for motorcoach design as priority items for consideration in future rulemaking, especially items on the National Transportation Safety Board (NTSB) "Most Wanted List" of safety improvements.

NHTSA asked the John A. Volpe National Transportation Systems Center (Volpe Center) to provide human factors research, evaluation, and technical support to assist in developing recommendations for updating emergency-evacuation-related requirements for motorcoaches currently contained in FMVSS 217.

This interim report describes the results of the first year of the two-year Volpe Center study and describes potential motorcoach design changes that may increase the egress rate of passengers and reduce the risk of injuries during emergency situations. While this effort is directed at intercity and charter / tour-over-the-road motorcoaches, insights and information considered relevant to emergency-egress requirements for other large buses or school buses are documented.

The three major topic areas addressed in this interim report are: 1) emergency exits, 2) interior and exterior emergency exit marking, and 3) emergency exit lighting.

To gain additional knowledge relating to various factors that affect motorcoach passenger egress rates using various emergency exit paths, Volpe Center staff completed the following activities during the first year of a two-year study:

- Conduct of literature search to review prior research on these three topics and to document how these matters are addressed by other nations.

- Naturalistic observations of passengers exiting from motorcoaches located at large bus terminals under normal conditions;

- Development of instrumentation to measure release and opening forces, primarily for emergency exit windows and emergency roof exit hatches;

- Field visits to:
    - Inspect current-design motorcoaches, in terms of front door, emergency exit window, and emergency roof exit hatch design and marking,
    - Complete force measurements for the latter two types of exits,
    - Operate and use emergency exit windows and roof emergency exit hatches on two different motorcoach models; and

- Human factors egress experiments to determine egress flow rate estimates for different types of exits, using:
  - Full bus of subjects for front door egress, and
  - Smaller number of subjects for emergency exit window and wheelchair-access-door egress.

OVERVIEW OF FINDINGS

Literature Review

The principal findings from the review are that:

- Very little research focused on bus emergency egress has been conducted since Federally funded work was completed at the University of Oklahoma Research Institute in the 1970s.

- None of the existing research literature addresses egress through emergency exits currently installed in motorcoaches, which have sill heights and window weights much greater than those of buses tested in the 1970s.

- FMVSS 217 emergency exit requirements are different for school buses in various aspects than for other types of buses. Each school bus is required to have at least one side door and emergency exit identification requirements. These requirements could be adapted for application to motorcoaches.

- Other U.S. transportation regulatory agencies specify requirements for emergency exits, including emergency exit identification and emergency lighting that could be adapted for application to motorcoaches.

- The Economic Commission for Europe (ECE) Commission has established standards for motorcoach emergency egress that could be adapted for application to motorcoaches. These standards include requirements for a second side-service or emergency door, larger emergency roof exit hatches than those required in the U.S., floor exit hatches, identification of emergency exits and instructions for their operation on the bus exterior, etc.

Egress Rates

Naturalistic observations of normal front-door egress at bus terminals showed that it normally requires three to four minutes for all passengers to egress from a fully loaded motorcoach, i.e., the egress rate is less than 20 passengers per minute (ppm). However, occasional clusters of passengers with minimal hand luggage were observed exiting at rates above 30 ppm.

Release and opening forces for U.S. motorcoach emergency exit windows are within current FMVSS 217 regulatory limits; however, the margin is small for the larger exit windows. The opening force for top-hinged emergency-exit windows (currently used on all U.S. motorcoaches) increases with the size of the opening. Many adults require a larger opening size than the minimum specified in FMVSS 217 to egress safely than the current test procedure specifies.

A series of human factor experiment trials was conducted at the Volpe Center campus, located in Cambridge, MA, using a motorcoach with federal employees as volunteer subjects. These trial results generated-egress time estimates using the front-door, emergency-exit windows, and wheelchair-access-door, all under ideal daytime conditions; the tests started with the egress path already locked in the open position, as shown in the following table.

Volpe Center Preliminary Motorcoach Egress Estimate – 56 Passengers

EGRESS PATH	NUMBER OF EXITS. USED	OPENING TIME (min)	FLOW RATE (ppm)	EGRESS (min)	TOTAL (min)
Front door	1	.05	36	1.56	1.61
Windows	6	.2	9	1	1.20
Wheelchair-access door	1	.2	25	2.24	2.44
Roof hatch	2	.1	12	2.33	2.43

All of the estimates are based on the assumption that all passengers know how to use the exits and have "hold open" devices for the windows and are based on the behavior of volunteers who judged themselves to be capable of performing the required actions without risk of injury. The flow rate (ppm) estimates for emergency roof exit hatches were derived from observations of only a very limited number of federal and contractor employees at the MGA Test Facility, located in Burlington, WI, using two motorcoaches overturned during NHTSA-conducted rollover testing.

The forces required to release and open roof emergency roof exit hatches were typically less than one half of the current regulatory limit value, 268 N (60 lbs), when measured with the bus upright. With the bus overturned on its side, roof hatch release and opening forces were negligible. The majority of able-bodied adults can egress through the emergency roof exit hatch of an overturned bus at the rate of approximately 12 ppm. Individuals of more limited physical ability can each take a minute or more to pass through the exit hatch, unless they are assisted by other passengers, or the bus driver.

These results indicate that timely motorcoach passenger egress could be achieved using any of these exit locations. However, several obstacles may exist during an actual motorcoach emergency:

- In a frontal motorcoach crash, the front door may be blocked, or the driver may be incapacitated. Without a driver to operate the door control, passengers may incur substantial delay in figuring out how to open the door.

- Passengers who try to use the emergency exit windows may find it difficult to raise and maintain them at sufficient height to allow rapid egress.

- Wheelchair-access doors cannot be opened from inside of any of the motorcoaches currently in use.

- Emergency roof exit hatches are useful only when a bus is overturned.

Emergency Exit Identification

The Volpe Center-conducted field measurements of illumination and letter sizes for motorcoach emergency exit signage showed that those in current use comply with FMVSS 217 requirements, whenever there is at least a low level of illumination present (i.e., in daylight or when the fluorescent boarding lights or adjacent reading lights are in use). However, the typical level of illumination provided during the night, 0.1 to 0.8 lux (0.01 to 0.08 foot-candle), by the "night lights") does not allow some exit signage to be conspicuous or easily legible, even at very short viewing distances.

SUMMARY OF PRELIMINARY DESIGN CHANGE CONSIDERATIONS

Issues relating to barriers to rapid motorcoach emergency egress have been documented in various NTSB reports and previously completed NHTSA-funded research study reports. The results of the Volpe Center study conducted to date are consistent with the findings contained in those reports.

Other U.S. transportation regulatory agency requirements for vehicle emergency exits, including exit identification, and emergency lighting, could be adapted for application to motorcoaches. These requirements (extensively described in the Year 1 interim report) include: more than one emergency exit door, larger emergency roof exit hatches, photoluminescent marking of emergency exits on the interior, retroreflective marking of emergency exits on the exterior, and independently-powered emergency lighting.

Certain provisions of existing FMVSS 217 requirements for school bus emergency exits and standards established by the Economic Commission for Europe for motorcoaches operated in other countries could also be adapted and applied to motorcoaches.

Potential motorcoach design changes that may increase the passenger egress rate and reduce the risk of injuries during emergencies are:

- Positive "hold open" devices for:
    o Doors that can be used for emergency egress,
    o Emergency exit windows, and
    o Emergency roof exit hatches;
- Improved emergency exit identification:
    o Interior
        – Increase signage conspicuity,
            * Place signage and instructions in location on or near the top or side of the exit that are more visible and of a color that contrasts with its background,
            * Use signs and instructions with a larger minimum specific letter height, and
            * Use:
                ~ "High-performance photoluminescent" material,
                ~ Illumination by emergency lighting powered by crash-survivable, self-contained independent power sources, or
                ~ Dual-mode signs (that combine both technologies), and
        – Provide clearer, more easily understood instructions for passengers to release and open:
            * Front door, for emergency egress, if the driver is incapacitated, and
            * Roof exit hatch, when the bus is overturned, and
    o Exterior
        – Provide retroreflective signage and markings to identify location of emergency door exits, and emergency roof exit hatches, and
        – Provide instructions for opening emergency door exits and emergency roof exit hatches;
- Increased minimum number and size of emergency roof exit hatches:
    o At least two hatches per motorcoach, and
    o Larger aperture dimension of 4,000 cm^2 (620 in^2); and
- An additional floor-level door exit on the side, located in the middle / rear half of the bus for passenger egress in an emergency, by either:

- o Modification of the wheelchair-access door to permit it to be opened from inside for use as an emergency exit, and /or
- o Another side door exit that could be opened and used as an emergency exit.

FURTHER RESEARCH

Volpe Center staff are investigating the following motorcoach emergency egress topic items in Year 2 of this study:

- Conduct of human factors experiments to determine:
  - o Rates of egress through:
    - Wheelchair-access doors with two different configurations and clearances, and
    - Stairways with 30 cm (12 in) step risers similar to those used for the second service door on many buses operated in other countries;
  - o Effects of illuminance levels on egress rates, and
  - o Human strength of adult subjects (evenly balanced by age and gender) to apply the pulling and pushing forces needed to open doors and top-hinged emergency exit windows; and
- Development of potential performance specifications and criteria for emergency exit identification using:
  - o Photoluminescent materials, including luminance, dimensions, and contrast requirements,
  - o Electrically powered-illuminated devices, and
  - o Dual-mode systems, combining both technologies.

TABLE OF CONTENTS

Section   **Page**

EXECUTIVE SUMMARY vii

1. INTRODUCTION 1

    1.1 Background 2
        1.1.1 U.S. Motorcoach Regulations 2
        1.1.2 NTSB Recommendations 3

    1.2 Purpose and Objective 4
    1.3 Scope 4
    1.4 Study Approach 5
    1.5 Report Organization 5

2. LITERATURE REVIEW 7

    2.1 NHTSA and FMCSA Requirements for Buses 7
        2.1.1 NHTSA and FMCSA Requirements for Buses (Motorcoach) and School Buses 7
        2.1.2 Historical Background 16
        2.1.3 Recent NHTSA Regulatory-Related Activities 21
        2.1.4 University of Oklahoma Research Institute Studies 22
        2.1.5 FMCSA Pre-Trip Safety Briefing Guidance 29

    2.2 Other U.S. Transportation Vehicle Requirements 30
        2.2.1 FAA 30
        2.2.2 FRA 30
        2.2.3 APTA PRESS Standards 31
        2.2.4 USCG 31

    2.3 International Bus Regulations 32
        2.3.1 Economic Commission for Europe (ECE) 32
        2.3.2 Canada 32
        2.3.3 Australia 32
        2.3.4 United Kingdom 33

    2.4 Other Relevant Information 33
        2.4.1 NTSB Reports 33
        2.4.2 1983 NBS Technical Note 1180 34
        2.4.3 Human Engineering Guide to Equipment Design 34
        2.4.4 MIL-STD 1472F Design Criteria Standard for Human Engineering 34
        2.4.5 UK Strength Data – 2002 34
        2.4.6 TCRP 100 Manual 34
        2.4.7 Volpe Center Alaska Bus Egress Study 35

    2.5 Other 35
    2.6 Summary 35

TABLE OF CONTENTS (cont.)

Section   **Page**

3. MOTORCOACH EGRESS HUMAN FACTORS EXPERIMENTS 37

   3.1 Naturalistic Passenger Observations 37

      3.1.1 Methodology 38
      3.1.2 Results 39

   3.2 Exit Opening and Release Force Measurements 39

      3.2.1 Existing FMVSS 217 Requirements 39
      3.2.2 Field Measurements 39

   3.3 MGA Test Facility Field Visit 46
   3.4 Volpe Center Egress Experiments 47

      3.4.1 Overview 47
      3.4.2 Experiment Study Design 48
      3.4.3 Subject Selection 50
      3.4.4 Experiment Trial Protocols 51

   3.5 Emergency Roof Exit Hatch Egress Experiments 53

      3.5.1 Somersault 54
      3.5.2 Whole Body Lift 54
      3.5.3 Cautious Approach 54

   3.6 Results 54

4. FRONT DOOR EGRESS 55

   4.1 FMVSS 217 Requirements 55
   4.2 Designs in Use 56
   4.3 Usability and Egress Rates 59

      4.3.1 Naturalistic Observation of Egress Flow Rate Results 60
      4.3.2 Volpe Center Experiment Front Door Egress Flow Rate Results 65

   4.4 Requirements from Other U.S. Modes 66
   4.5 International Bus Requirements 67
   4.6 Relevant Research 68
   4.7 Discussion 69
   4.8 Considerations 70

5. WHEELCHAIR-ACCESS DOOR AND OTHER SIDE DOOR EGRESS 73

   5.1 FMVSS 217 Requirements 73
   5.2 Designs in Use 75
   5.3 Usability and Egress Rates 78

      5.3.1 Movement of Seats Away from Wheelchair-Access Door 78
      5.3.2 Wheelchair-Access Door Experiment Egress Flow Rate Results 79

TABLE OF CONTENTS (cont.)

Section | **Page**

    5.4   Requirements from Other U.S Modes 79
    5.5   International Bus Requirements 80
    5.6   Relevant Research 81
    5.7   Discusssion 82
    5.8   Considerations 84

6. EMERGENCY EXIT WINDOW EGRESS 87

    6.1   FMVSS 217 Requirements 87
    6.2   Designs in Use 89
            6.2.1   Motorcoaches 89
            6.2.2   School and Transit Buses 92

    6.3   Ability and Egress Rates 93
            6.3.1   Volpe Center Emergency Exit Window-Release Force Measurements 93
            6.3.2   Volpe Center Emergency Exit Window Opening Force Measurements 94
            6.3.3   Volpe Center Emergency Exit Window Experiment Egress Flow Rate Results 95

    6.4   Requirements From Other U.S. Modes 96
    6.5   International Bus Requirements 97
    6.6   Relevant Research 98
    6.7   Discussion 101
    6.8   Considerations 102

7. EMERGENCY ROOF EXIT EGRESS 103

    7.1   FMVSS 217 Requirements 103
    7.2   Designs in Use 103
    7.3   Usability and Egress Rates 106
    7.4   Requirements From Other U.S. Modes 107
    7.5   International Bus Requirements 107
    7.6   Relevant Research 108
    7.7   Discussion 109
    7.8   Considerations 110

8. EMERGENCY EXIT IDENTIFICATION 111

    8.1   FMVSS 217 Requirements 111
    8.2   Designs in Use 112
    8.3   Requirements From Other Modes 117
    8.4   International Bus Requirements 120
    8.5   Relevant Research 120
    8.6   Discussion 122
    8.7   Considerations 124

TABLE OF CONTENTS (cont.)

Section	Page
9. EMERGENCY EXIT LIGHTING	**125**
9.1 FMVSS 217 Requirements	125
9.2 Designs in Use	126
9.3 Requirements from Other Modes	128
9.4 International Bus Requirements	131
9.5 Relevant Research	132
9.6 Discussion	133
9.7 Considerations	136
10. FINDINGS, CONSIDERATIONS, AND FURTHER RESEARCH	**137**
10.1 Summary of Findings	137
10.2 Summary of Preliminary Design Change Considerations	139
10.3 Further Research	140
11. REFERENCES	**143**
APPENDIX A. Bus Emergency Egress Regulation History	A-1
APPENDIX B FMVSS 217 Bus Egress Requirements	B-1
APPENDIX C NHTSA Motorcoach / School Bus and Other Vehicle Egress Design Regulations	C-1
APPENDIX D. FMCSA Bus / Motorcoach Safety Brochure – Lift and Pull Windows	D-1

LIST OF FIGURES

Figure	Page
Figure 3-1. Fingertip Release – Emergency Roof Exit Hatch	41
Figure 3-2. Load Cell and Free-Body Diagram of Window-Opening Forces	41
Figure 3-3. Force Measurement Graph – Peak Release Force of 122.9 N	42
Figure 3-4. Force Required to Open Emergency Exit Window to a Specified Angle	44
Figure 3-5. Force Gauge Attached to an Extension	45
Figure 3-6. Motorcoach Emergency Exit Window Open with Force Gauge	46
Figure 3-7. Volpe Egress Experiment Motorcoach	48
Figure 3-8. Volpe Center Motorcoach Experiment Configuration and Camera Location	49
Figure 3-9. Pictogram Illustrating the Sitting Jump Egress Method	52
Figure 4-1. Motorcoach Front Door and Stairs	56
Figure 4-2. School Bus Front Door and Stairs	56
Figure 4-3. Transit Bus Front Door and Stairs	57
Figure 4-4. Motorcoach Interior Latches for Front Door	57
Figure 4-5. Door Pneumatic Actuator Pressure Release Knob and Instructions	58
Figure 4-6. Hand-Crank Front Door Opener	58
Figure 4-7. Motorcoach Driver Front Door Controls	59

LIST OF FIGURES (cont.)

Figure		Page
Figure 4-8.	School Bus and Transit Bus Driver Front Door Controls	59
Figure 4-9.	Volpe Center Motorcoach Experiment to Determine Front Door Egress Rate – Best-Case	66
Figure 4-10.	Volpe Center Commuter Rail Experiment to Measure Normal Egress Rate	68
Figure 5-1.	FMVSS 217 Clearance Requirements for School Bus Side Door Emergency Exits	74
Figure 5-2.	FMVSS 217 School Bus Wheelchair Anchorage Location Prohibition	74
Figure 5-3.	Motorcoach Wheelchair-Access Door Location Interior View – Examples	75
Figure 5-4.	Motorcoach Wheelchair-Access Door / Lift Exterior View – Examples	76
Figure 5-5.	School Bus Wheelchair-Access Door Emergency Exit – Closed Position	76
Figure 5-6.	School Bus Wheelchair-Access Door – Closed Position	76
Figure 5-7.	Transit Bus Rear Side Wheelchair-Access Door Location – Example	77
Figure 5-8.	Seat Configuration with a 25.4-cm (10-in) Clearance (Aircraft) – Example	77
Figure 5-9.	Volpe Center Subjects – Sitting Jump Using Wheelchair-Access Door	79
Figure 5-10.	MAN Motorcoach – Middle Side Service Door	80
Figure 5-11.	ECE 36 – Access to Emergency Doors (Annex 3, Figure 2)	81
Figure 5-12.	Subjects Jumping from *GM* PD4104 Bus Rear Emergency Exit (OKRI 1972)	82
Figure 6-1.	Ellipsoids Formed by Rotation of 50 by 33 cm (20 by 13 inch) Ellipse	88
Figure 6-2.	Subject Egress Using 50 by 33 cm (20 by 13 in) Elliptical Opening (OKRI 1970)	89
Figure 6-3.	*MCI* Emergency Exit Window Release Bars	90
Figure 6-4.	Close-Up View of *MCI* Sill-Mounted Emergency Exit Window Release Components	90
Figure 6-5.	*Prevost* Sash-Mounted Emergency Exit Window Release Bars	91
Figure 6-6.	*Prevost* Sill-Mounted Emergency Exit Window Latch Components	91
Figure 6-7.	*Van Hool* Sash-Mounted Emergency Exit Window Release Mechanism	92
Figure 6-8.	*Van Hool* Emergency Exit Window Sill-Mounted Release Components	92
Figure 6-9.	School and Transit Bus Side Emergency Window Exits -- Examples	93
Figure 6-10.	Subject Performing Motorcoach Emergency Exit Window Egress	96
Figure 6-11.	Motorcoach Emergency Exit Window Egress – +95th Height Percentile Male	96
Figure 6-12.	1972 OKRI Intercity Bus Egress Experiments (left and right side)	98
Figure 6-13.	1984 Hungarian Bus Window Egress Tests	99
Figure 6-14.	Test Apparatus for Measuring Finger-Tip Pulling Strength	100
Figure 6-15.	UK Subject Distribution – Maximum Finger-Tip Pulling Strength	100
Figure 7-1.	*MCI* Emergency Roof Exit Hatches – Interior	104
Figure 7-2.	*Prevost* and *Van Hool* Emergency Roof Exit Hatch	104
Figure 7-3.	*MCI* Open Emergency Roof Exit Hatch – Exterior	105
Figure 7-4.	School Bus Emergency Roof Exit Hatch	105
Figure 7-5.	School Children Crawling through Roof Hatch Mockups (OKRI 1972)	109
Figure 8-1.	Motorcoach Front-Door Operation Instructional Signage Example – Interior	113
Figure 8-2.	School Bus Front-Door Instructional Signage Example – Interior	113
Figure 8-3.	Motorcoach Emergency Exit Window Signs and Instructions (1)	114
Figure 8-4.	Motorcoach Emergency Exit Window Signs and Instructions (2)	114

LIST OF FIGURES (cont.)

Figure — **Page**

Figure 8-5. Motorcoach Emergency Exit Window Identification, Instructions, and Sign 115
Figure 8-6. School Bus Window Emergency Exit Signage and Instructions 115
Figure 8-7. Transit Bus Door and Window Emergency Exit Signage – Interior 116
Figure 8-8. Motorcoach and School Bus Emergency Roof Exit Hatch – Interior Signage and Instructions 116
Figure 8-9. School Bus Emergency Exit – Exterior Retroreflective Marking 117
Figure 8-10. Passenger Rail Car – Photoluminescent Emergency Exit Signage 118
Figure 8-11. Illustration of Off-Axis Sign Viewing 121
Figure 8-12. Letter-Size Requirements as a Function of Distance and Luminance 122
Figure 9-1. Typical Motorcoach – Bright Fluorescent Lighting 127
Figure 9-2. Typical Motorcoach – Reading Lights On 127
Figure 9-3. Outline Lighting on Bus Step 128
Figure 9-4. Aircraft Electrically-Illuminated Emergency Exit Sign 129
Figure 9-5. Aircraft Floor Proximity Exit Path Marking – Electrically Powered 130
Figure 9-6. Aircraft Floor Proximity Exit Path Marking – Photoluminescent 130
Figure 9-7. LED Emergency Light with Capacitive Energy Storage 135

LIST OF TABLES

Table — **Page**

Table 2-1. 1970 OKRI Subject Egress Rates for Escape Through Various Size Ellipses 24
Table 2-2. Summary of 1972 OKRI Intercity Bus Egress Rates 26
Table 2-3. Summary of 1978 OKRI Intercity Bus Egress Rates 28
Table 3-1. Emergency Exit Window Dimensions and Weight 44
Table 4-1. Observed Motorcoach Passenger Egress – South Station, Boston, MA, December 27, 2007 61
Table 4-2. Observed Motorcoach Passenger Egress – South Station, Boston, MA, May 23, 2008 62
Table 4-3. Observed Motorcoach Passenger Egress – Mendenhall Glacier Visitors Center, Juneau, AK, August 14-15, 2006 (1) 63
Table 4-4. Volpe Center Motorcoach Front-Door Egress Experiment Results 66
Table 4-5. Summary Observations of Front-Door Egress Rates 70
Table 5-1. Wheelchair-Access Door Row Clearances for a *Prevost X-345* Bus 78
Table 6-1. Motorcoach Emergency Exit Window – Peak Release Force Measurements 94
Table 6-2. Motorcoach Emergency Exit Window Opening Forces – Actual Measurements 95
Table 7-1. Volpe Center Emergency Roof Exit Hatch Release Force Measurements 106
Table 10-1. Volpe Center Preliminary Motorcoach Egress Estimate – 56 Passengers 138

1. INTRODUCTION

The mission of the National Highway Traffic Safety Administration (NHTSA), United States Department of Transportation (USDOT), is to reduce motor vehicle crashes and injuries. NHTSA safety-related regulations for buses and school buses are included in extensive Federal Motor Vehicle Safety Standards (FMVSS), as contained in Title 49, Code of Federal Regulations (49 CFR), *Part 571*.[1]

Subsection 571.217 (FMVSS 217), *Bus Emergency Exits and Window Retention and Release*[2] specifies a series of release and retention tests for all windows, other than windshields; as well as a series of dimensional and physical requirements for bus emergency exits, including their size, location, opening forces, and marking. The intent of FMVSS 217 is: "to minimize the likelihood of occupants being ejected from the bus and to provide a means of readily accessible emergency egress" for those occupants under a variety of bus crash and other emergency scenarios. These scenarios can include catastrophic crash or other emergency situations, such as a fire, rollover, or water immersion, where immediate emergency egress is necessary under life-threatening and hazardous conditions.

In 2007, NHTSA prepared a comprehensive research plan to address motorcoach[*] safety issues, identifying several improvements for motorcoach design as priority items for consideration in future rulemaking.[3] One consideration identified in this plan is to address items on the National Transportation Safety Board (NTSB) "Most Wanted List" of safety improvements, i.e., "easy opening windows and roof hatches that stay open during evacuations" (H-99-9).[4] (Other NTSB recommendations to NHTSA relating to bus emergency egress are listed in Subsection 1.1.2)

NHTSA asked the Volpe National Transportation Systems Center (Volpe Center), Research and Innovative Technology Administration (RITA), USDOT, to provide human factors research, evaluation, and technical support, to assist the agency in developing recommendations for updating the large bus emergency egress-related requirements for large buses, as currently contained in FMVSS 217.

This interim report describes preliminary findings and topics for NHTSA consideration, including potential design changes that may increased during motorcoach emergency egress, as developed during the first year of a two-year study. A report containing final results and topics for NHTSA consideration will be published after the Volpe Center completes the remaining study activities in 2009.

---

[*] The NHTSA research program plan refers to motorcoaches as "intercity-transport buses."

1.1 BACKGROUND

NHTSA defines buses as "motor vehicles with motive power, except a trailer designed for carrying more than 10 persons".[5] (This includes the driver.) NHTSA defines "school bus" as a bus that is "sold or introduced in interstate commerce for purposes that include carrying students to and from school or related events but does not include a bus designed and sold for operation as a common carrier in urban transportation."

1.1.1 U.S. Motorcoach Regulations

The NHTSA regulatory focus of the FMVSS 217 requirements is bus window retention and emergency exit safety performance criteria, which apply to all buses (with the exception of buses used to transport prisoners), including large buses, such as motorcoaches and school buses.

Accordingly, in addition to motorcoaches operated in intercity, tour, commuter, and charter service, NHTSA regulations also apply to all buses manufactured and sold for use on highways, including transit buses operated by cities and towns and school buses.

Actual requirements vary according to the type of bus and the gross vehicle weight rate (GVWR). (See Chapter 2.)

Emergency exit requirements are different for school buses in various ways than for other types of buses. FMVSS 217 allows motorcoach and other bus manufacturers to follow school bus requirements for the number and type of emergency exits, rather than non-school bus requirements. However, NHTSA is not aware of any manufacturers that build buses to those school bus requirements.

FMVSS 217 bus safety requirements are organized by topic area:

- Window retention,
- Provision of emergency exits,
- Emergency exit release,
- Emergency exit opening,
- Emergency exit identification, and
- Test conditions.

Under most emergency circumstances, passengers simply get off the bus through the front service door after the driver pulls the bus off the road to a safe location. However, in an emergency that involves a crash, FMVSS 217 requires a distribution of emergency exit windows and emergency roof hatch exit(s) that can be used if the front door is not available or if the bus is on its side.

The Federal Motor Carrier Safety Administration (FMCSA) also has jurisdiction over buses that operate in interstate commerce.[6] FMCSA regulations cover operations, and their regulatory focus is on the companies who actually operate motorcoach service. Specific NHTSA and FMCSA motorcoach emergency egress-related regulations are reviewed in greater detail in Chapter 2.

1.1.2 NTSB Recommendations

As a result of its investigation of the 2005 Wilmer, TX bus fire[7] and recent motorcoach crashes,[8,9,10] the NTSB issued several recommendations to NHTSA pertaining to emergency exit designs, emergency lighting using self-contained power, and photoluminescent and retroreflective signs and marking for emergency exit identification:

- Evaluate current emergency evacuation designs of motorcoaches and buses by conducting simulation studies and evacuation drills that take into account, at a minimum, acceptable egress times for various emergency situation environments, including fire and smoke; unavailable exit situations; and the current above-ground height and design of window exits to be used in emergencies by all potential vehicle occupants. (HS-07-8)

- Revise FMVSS 217 to require:
  - That all motorcoaches be equipped with emergency lighting fixtures that are outfitted with a self-contained independent power source. (H-00-01)
  - The use of interior luminescent or exterior retro-reflective material, or both, to mark all emergency exits in all motorcoaches. (H-00-02)
  - That other than floor-level emergency exits can be easily opened and remain open during an emergency evacuation when a motorcoach is upright or at unusual attitudes. (H-99-9)

In addition, NTSB issued two other recommendations to the USDOT that guidance be developed for motorcoach pre-trip safety briefings to include information to passengers similar to that provided for commercial airline passengers (HS-99-7 and H-99-8).[8] (In response, the Secretary of Transportation directed that FMCSA to develop that guidance. FMCSA has developed and issued that guidance, which is described in Subsection 2.1.5.)

1.2 PURPOSE AND OBJECTIVE

The three topic areas that NHTSA selected for research in its research plan[3] include:

1) Emergency egress, e.g., the number and type of emergency exits;
2) Signage, including the use of photoluminescent materials; and
3) Illumination, e.g., emergency lighting.

NHTSA requested that the Volpe Center undertake a research effort to identify human factors issues related to motorcoach emergency egress. This request was in recognition of the Volpe Center's prior and ongoing emergency egress studies for other USDOT modal administrations.

The purpose of the study is to provide human factors research, evaluation, and technical support to the NHTSA Vehicle Safety Research Office, which will develop and implement strategies to enhance passenger emergency egress from motorcoaches.

The objectives of this report are to:

- Address issues raised by the NTSB and the NHTSA research plan; and
- Identify potential motorcoach design changes that may improve passenger safety during emergency egress.

1.3 SCOPE

This interim report describes the results of the Volpe Center first-year tasks of a two-year study. The report topics are focused on emergency egress requirements applicable to motorcoaches and other large buses, other than school buses operated in the U.S. Intercity and charter/tour buses are large buses, e.g., "motorcoaches" that travel "over the road" and make infrequent stops between cities. While the study is directed at intercity and charter / tour motorcoaches, insights and information considered relevant to emergency egress requirements for other buses or school buses are also documented.

The three major topic areas addressed in this interim report are: 1) emergency exits, 2) interior and exterior emergency exit marking, and 3) emergency exit lighting. (This study does not address emergency exit window or other window retention during motorcoach crash impacts.)

The human factors issues in each of these topic areas are discussed in the context of ambulatory, non-disabled adult passengers. All emergency exits in buildings, as well as passenger aircraft, ships, and trains, are designed for ambulatory persons. In an emergency, able-bodied passengers are expected to release and open the emergency exits and assist other passengers who require help.

1.4 STUDY APPROACH

Volpe Center staff worked with NHTSA staff to address NHTSA priority action items for motorcoaches. The work is being conducted during a two-year time period that began in October 2007.

Volpe Center activities were grouped into several major areas. The first general area includes the review and evaluation of existing U.S., international, and selected country motorcoach regulations and relevant research, as well as the identification of common issues related to motorcoach emergency evacuation. In addition, the Volpe Center collected information about current U.S. motorcoach emergency exit designs. The Volpe Center also identified, obtained, and reviewed emergency evacuation regulations and related research for other mass transportation vehicles, such as passenger aircraft, passenger trains, and passenger ships. The second general area includes field visits to the MGA Research Corporation (MGA) test facility, and three motorcoach operators. The third area includes Volpe Center-designed and conducted human factors experiments. These field visits and experiments ranged from measuring emergency exit window opening forces to determining egress flow rates from various types of exits (including the front door, side emergency exit window, wheelchair-access door, and emergency roof exit hatch).

Information from these activities was used to evaluate the potential applicability of other regulations, as well as the usability of alternative emergency exit designs, emergency lighting, and interior and exterior marking.

During Year 2, the Volpe Center is performing additional tasks in consultation with NHTSA, including further motorcoach bus egress and other experiments prior to developing final topic considerations for revising FMVSS 217.

1.5 REPORT ORGANIZATION

Chapter 2 summarizes existing U.S and international motorcoach and school bus emergency egress-related regulations, guidance, and research; related U.S. transportation emergency egress regulations; and related research.

Chapter 3 summarizes the human factors experiments conducted by Volpe Center staff using U.S. motorcoaches.

Chapters 4-7 discuss bus occupant emergency egress: front door egress (Chapter 4), wheelchair-access door egress (Chapter 5), emergency exit window egress (Chapter 6), and emergency roof exit hatch egress (Chapter 7). Chapter 8 discusses emergency exit identification, and Chapter 9 discusses emergency exit lighting.

Chapters 4-9 follow a similar format: applicable NHTSA (and if appropriate, FMCSA) regulations for motorcoaches and school buses are summarized, current designs are described; usability issues are identified; and other U.S. transportation regulatory agency and, as available, international motorcoach requirements are summarized. In addition, the information in those sections is discussed, preliminary considerations are described, and related Year 2 study activities are also identified.

Lastly, Chapter 10 summarizes Year 1 study findings and topic considerations for Chapter 2-9. A brief overview of Year 2 study tasks is also included.

2. LITERATURE REVIEW

This chapter summarizes the review of regulations and other information pertaining to transportation vehicle emergency egress, with specific focus on applicability to emergency egress from motorcoaches and other large buses. The review included:

- NHTSA and related FMCSA regulations, guidance, and research pertaining to bus (including motorcoach) and school bus emergency egress;
- International regulations, guidance, and research related to buses, with emphasis on motorcoaches;
- Other U.S. transportation vehicle regulations, guidance, and research;
- Pertinent other research; and
- Other information of interest.

2.1 NHTSA AND FMCSA REQUIREMENTS FOR BUSES

NHTSA regulations for bus and school bus design and manufacture are specified in 49 CFR, Part 571, *Federal Motor Vehicle Safety Standards* (FMVSS).[1] NHTSA defines "buses" as "a motor vehicle with motive power, except a trailer, designed for carrying more than 10 persons." "School bus" is defined as a bus that is "sold or introduced in interstate commerce for purposes that include carrying students to and from school or related events but does not include a bus designed and sold for operation as a common carrier in urban transportation."

49 CFR, Part 393, *Parts and Accessories*, and Part 396, *Inspection, Repair, and Maintenance* apply to buses operated in interstate commerce.[11]

Subsection 2.1.1 summarizes NHTSA and FMCSA requirements for motorcoach and school bus emergency egress, while Subsection 2.1.2 summarizes the historical background for those requirements. Subsection 2.1.3 summarizes recent NHTSA-regulatory-related activities.

2.1.1 NHTSA and FMCSA Requirements for Buses (Motorcoach) and School Buses

FMVSS 217 requirements apply to the safety performance of buses (with the exception of buses used to transport prisoners) and school bus window retention and emergency exits. FMVSS 217 provisions specify: 1) retention test requirements for all windows (including emergency exit windows); and 2) dimensional and physical requirements for emergency exits, including their size, location, opening and release operation and forces; and 3) emergency exit identification.

The intent of the requirements is "to minimize the likelihood of occupants being thrown from the bus and to provide a means of readily accessible emergency egress" to occupants under a variety of crash and other emergency scenarios. These bus emergency scenarios can include catastrophic situations, such as fires, rollovers, or water immersion, where immediate emergency evacuation is necessary under life-threatening and difficult conditions.

Prior to 1973, the Interstate Commerce Commission (ICC) and Federal Highway Department (later FMCSA) regulations for bus emergency exits applied only to buses (including motorcoaches) operated in interstate commerce.

In 1972, NHTSA issued FMVSS 217 requirements for bus window retention and emergency exits[12] that became effective in 1973, for buses, other than school buses. (However, if school buses were equipped with emergency exits, those exits were required to comply with FMVSS 217 requirements.) All bus side and rear push-out emergency exit windows were required to be retained in a manner to assure adequate resistance to outward forces. All large buses, including intercity and charter motorcoaches (and transit buses), were required to provide at least 67 square inches of emergency exit area per seating position, 40 percent of which had to be on each side of the bus. All buses, other than school buses, were also required to have push-out windows and a rear emergency door to provide unobstructed emergency exit area. If the bus construction precluded the installation of a rear emergency exit door, an emergency roof hatch exit was required.

Additional requirements for school bus emergency exits, including requirements for at least one side emergency exit door, or one rear emergency exit door with a push-out window, first became effective in 1977.

While NHTSA updated and expanded the school bus emergency exit requirements after publishing several proposed revisions to FMVSS 217, motorcoach emergency exit requirements have remained essentially unchanged, with exceptions for more stringent emergency exit markings and instructions, effective in 1976; and use of sliding emergency exit windows was permitted, effective in 1996.

While NHTSA modified FMVSS 217 in 1995 to permit non-school bus emergency exits to meet school bus requirements for the type and size of emergency exits, NHTSA is not aware of any bus manufacturers that have taken advantage of this alternative compliance option.

Lastly, FMVSS 220 requires school bus rollover protection (roof crush) tests, which include specifications that emergency exits remain operable during and after the tests.[13]

Appendix A contains a summary table that lists the time sequence of all FMVSS 217 (and FMVSS 220) revisions.

(Note: The rule descriptions contained herein are derived from the actual NHTSA and FMCSA regulations contained in 49 CFR and are for information only.

2.1.1.1 *FMVSS 217*

The following sections present a summary of FMVSS 217 requirements for window retention, provision of emergency exits, exit release, exit opening size, exit identification, and test conditions. (Major NHTSA regulation section numbers are provided for each topic for reference). FMVSS 217 requirements may vary according to the type of bus design and the GVWR. NHTSA specifies requirements for two values of GVWR for vehicles: 1) with more than 10,000 lbs, and 2) with 10,000 lbs or less. The focus of the following description is directed at regulations applicable to buses and school buses with GVWR of more than 10,000 lbs.

Appendix B contains a comparison of motorcoach bus and school bus requirements.

A. Window Retention (S5.1)

All bus and school bus windows must be retained in a manner that would assure adequate resistance to outward forces. Window retention requirement tests apply to all bus side and rear windows, including emergency exit windows installed on motorcoaches and school buses.

The test measures ability to retain the window glazing and each surrounding structure using an increasing force applied by a specified head form traveling at the rate of 5 cm (2 in) per minute, under specified test conditions. The required performance criteria is the prevention of any opening large enough to permit a 4-inch-diameter sphere, under a force including the weight of the sphere of 5 lbs (2 kg) until any one of three conditions are met. These conditions include: reaching 1,200-lbs force, shattering or cracking 80 percent of the window, or the inner surface of the window is moved a prescribed distance. Windows less than 20 cm (8 in) in size are exempt.

B. Provision of Emergency Exits (S5.2)

Unobstructed openings of a minimum size must be provided on both motorcoaches and school buses for use as emergency exits during an emergency evacuation. For motorcoaches, total emergency exit opening surface area is based on the number of designated seating positions. With the exception of a rear door (or roof exit), motorcoach emergency exits are not required to be of a specific type. In contrast, school buses are required to be equipped with additional specific types of emergency exits, based on seating capacity, in addition to a side or rear door emergency exit.

Motorcoaches (Buses)

The unobstructed opening surface area for motorcoach emergency exits must collectively amount to, in total square centimeters, at least 432 times the number of designated seating positions on the bus. At least 40 percent of the required surface area must be located on each side of the bus. No single emergency exit can be credited to be more than 3,458 cm^2 (536 in^2) of the total exit area.

Sliding or push-out emergency exit windows are permitted. It appears that only transit bus manufacturers have used the sliding emergency exit window option since NHTSA issued the 1995 final rule, while motorcoach manufacturers have not.

The unobstructed opening requirements may be met by providing side exits and at least one rear exit. Although the regulation does not explicitly state that front side doors can be considered to be emergency exits, NHTSA has indicated that if the front door or wheelchair-access side door design and operation complies with all requirements for emergency exits, it can be counted as part of the emergency exit opening size calculation area.

The rear door emergency exit must meet exit release, opening size, and identification requirements when the bus is upright or overturned on either side. If the configuration does not permit the installation of an accessible rear exit, the manufacturer must provide a roof exit located in the rear half of the bus that meets the release, opening size, and identification requirements for emergency exits.

School Buses

School bus unobstructed emergency exit opening requirements are subject to minimum type and size provisions, focusing on rear and side door emergency exits, but with additional emergency exits based on seating capacity.

School buses are permitted to meet the minimum number of emergency exit requirements by using one of two options:

1) One rear emergency door on the vehicle that opens outward, hinged on the right side; or
2) One emergency door on the vehicle left side that opens outward, hinged on the forward side; with one push out rear window, with minimum opening of 41-cm (104-in) high and 122-cm (310-in) wide.

Regardless of whether Option 1 or 2 is chosen, school buses must have additional emergency exits (e.g., doors, emergency exit windows, and emergency roof exit hatches) based on capacity. These additional emergency exits may be installed in combinations, at the option of the manufacturer, as specified in several tables contained in FMVSS 217.

In addition, the area of a school bus door opening equipped with a wheelchair-access lift can be counted towards the additional necessary emergency exits, if it meets the requirements of Option 1 or 2 above (i.e., rear- or left-side door) and the lift folds or stows in such a manner that the area is available for use by persons not needing the lift. In addition, the lift must be manually operated and not require electrical power to operate. When the manual lift is in a folded or stowed position, the opening is considered to be an emergency exit.

If additional side emergency exit doors are used to comply with the capacity-based emergency exit requirements, they must meet the emergency door exit requirements and other location requirements, including separation of exits. Roof hatches may be installed to meet the additional exit requirements that are required for increased bus capacity. The hatches must be hinged on their forward side; operable from both the interior and exterior; and located either at the midpoint of the bus, if one hatch is installed, or located equidistant from each other, if more than one is installed.

Emergency exit windows must be of an even number, evenly divided between the left and right side, if they are used to meet the requirement for additional exits for increased bus capacity. Horizontal sliding emergency exit windows must not be used in order to minimize occupants putting their heads or arms out windows or throwing items out. To avoid confusion by children trying to decide how to open a particular emergency exit window, sliding and push-out window exits are not permitted on the same bus, with the exception of buses that use a single rear push-out emergency exit window and the remainder sliding side emergency exit windows.

The school bus engine starting system must not operate if any of the emergency exits are locked from either the inside or outside the bus. Locked means that the release mechanism cannot be activated at the door without a special device, such as a key or special information, as in a combination.

School buses manufactured after 1994 must comply with more stringent regulations, which require additional doors, emergency exit windows, and roof hatches, based on capacity.

C. Emergency Exit Release (S.5.3)

Motorcoaches and school bus emergency exit releases have several common requirements, including testing. Exit releases must allow for the manual release of the emergency exit, as tested within specified regions, by a single occupant, both before and after the required window retention test, using specified test conditions, under specified high- or low-force applications. FMVSS 217 includes several figures to define the boundaries of the high-force regions that correspond to regions in which an able-bodied person can exert maximum force, i.e., relatively close to shoulder height of an average male. FMVSS 217 also includes figures that define low-force regions, i.e., those that require reaching well below or well above occupant shoulder height.

Each exit must have a maximum of two release mechanisms. If one release mechanism is installed, two force applications must be used to release (and open) the exit. One of the force applications for each exit must differ from the direction of the initial motion to open the exit by not less than 90° and no more than 180°.

There are specific provisions for emergency exit release by bus type for each type of emergency exit.

Motorcoaches (Buses)

Release mechanisms must be located within specified regions, with the lower edge of the regions either 13-cm (5-in) above the adjacent seat or 5-cm (2in) above the arm rest, whichever is higher. Testing is required to determine that the forces required to release and open the exits do not exceed 268 N (60 lbs) in high-force regions and 89 N (20 lbs) in low-force regions.

School Buses

Requirements for school buses are more stringent than those currently applicable to motorcoaches in several respects.

The maximum permissible forces for releasing and opening all school bus exits in high-force regions are 178 N (40 lbs); in low-force regions, the force limits are the same as motorcoaches– 89 N (20 lbs).

Emergency exit door release mechanisms must allow release of the door both inside and outside the passenger compartment. The force application must be an upward motion from inside the bus and, at the option of the manufacturer, outside the bus. Each emergency exit door must be equipped with a positive door-opening device, which bears the weight of the door and keeps the door from closing regardless of bus body orientation.

Emergency exit door releases must also operate without the use of remote controls or tools, notwithstanding the failure of the vehicle power system. A pull-type mechanism can only be used if it is recessed and the handle does not protrude beyond the recessed regions. If the release is not in position to cause the door to be closed and the ignition is on, the school bus must provide continuous warning to the driver, and in the vicinity of the emergency exit. The engine starting system of a bus must not operate if any emergency exit is locked (requiring a key or combination to release) from either the inside or the outside of the bus.

Emergency exit window releases may be released by rotary or straight motions if located in low-force regions. Only straight motion perpendicular to the window surface is permitted in high-force regions, i.e., windows must push out. If the release is not in position to cause the window to be closed and the ignition is on, the bus must provide the same type of warning as described for emergency door exits.

Emergency roof exit hatch release mechanisms must be releasable from both inside and outside the passenger compartment. Rotary or straight motions are permitted if the release is located in a low-force region. In high-force regions, only force applications that are straight and perpendicular to the surface of the exit are permitted.

D. Emergency Exit Opening Size (S5.4)

Motorcoach and school bus emergency-exit openings must meet minimum size requirements. These requirements apply before and after the required retention test, under specified conditions, when manually opened by a single occupant under specified respective release conditions, including reach distances and corresponding forces. The openings must be large enough to allow an unobstructed passage of a "test fixture" of specified dimensions, which vary according to the type of bus.

Motorcoaches (Buses)

The size of the required emergency exit opening is based on the use of an ellipsoidal test fixture. The dimensions of the test fixture are generated by rotating a 50-by-33 cm (20-by-13-in) ellipse around its minor axis. Emergency exit openings must be large enough to permit passage of this fixture while keeping a major axis horizontal, i.e., the test fixture may be tipped.

School Buses

The unobstructed opening for rear emergency exit doors must be large enough to permit passage of a rectangular parallelepiped 1,145-mm (45-in) high, 610-mm (24-in) wide, and 305-mm (12-in) deep, keeping the 1,145-mm dimension vertical, the 610-mm dimension parallel to the opening, and the lower surface in contact with the bus floor at all times. The bottom edge of the rear-most surface of the parallelepiped must be tangent to the plane of the door opening.

The opening size for school bus side emergency exit doors must be at least 114-cm (44.8-in) high by 61-cm (24-in) wide. In addition, access to side doors must be provided by an aisle at least 30-cm (12-in) wide measured from the rear edge of the door (as shown in Figures 5A and 5 of FMVSS 217).

Emergency exit doors must be equipped with a positive door opening device that bears the door weight, keeps the door from closing past a specified point, and provides a release or override, regardless of the bus body orientation. No additional action must be required beyond opening the door past the point at which the door is perpendicular to the bus body.

Roof hatches must operate using specified force levels that permit an opening of at least 41 by 41 cm (16 by 16 in).

E. Emergency Exit Identification (S.5.5)

Each bus emergency exit must be identified by signs and concise operating instructions for unlatching and opening. There are specific requirements for motorcoaches and school buses.

Motorcoaches (Buses)

Each motorcoach emergency exit door must be designated by the words "Emergency Exit Door" or "Emergency Exit." Other emergency exits must be designated by the words "Emergency Exit." The exit markings and instructions must be located within 16 cm (6 in) of the release. Emergency exit markings must be legible to persons with 20/40 vision when normal nighttime light is available in the interior, by a person sitting in an adjacent seat or a person standing in the aisle location that is closest to the seat. The emergency exit markings must be legible at the seating area and other specified locations, when the seating area is occupied.

When the emergency exit release mechanism is not located within the occupant space of an adjacent seat, a label must be placed within the occupant space that indicates the location of the nearest emergency exit release.

School Buses

Each school bus emergency exit must be designated by the words "Emergency Exit Door" or "Emergency Exit," as appropriate. The letters must be at least 5-cm (2-in) high and be of a color that contrasts with the background they are applied to.

The exit designation for each emergency exit door and emergency exit window must be located at the top or directly above the exit (the window marking may be at the bottom of the window), on both the inside and outside surface of the bus. Emergency roof exit marking must be located on the surface of the exit or within 30 cm (12 in) of the exit, inside the bus, with operating instructions located within 15 cm (6 in) of the roof exit opening.

Concise operating instructions must be located on the inside of the bus, within 30 cm (12 in) of each exit, that describes the motions necessary to unlatch and open each emergency exit, with letters at least 1-cm (3/8 in) high, and of a color that contrasts with its background.

Each emergency exit opening must have its outside perimeter outlined with retroreflective[**] tape, with a minimum width of 2.5 cm (1 in) that is red, white, or yellow in color. The tape retroreflectivity must meet specified minimum criteria,[***] as contained in Table 1 of FMVSS 217 under the test conditions required in S6.1 of FMVSS 125, Warning Devices.[15]

For school buses equipped with one or more wheelchair anchorage positions, labels must be provided on the inside surface directly beneath or above each required "Emergency Exit Door" or "Emergency Exit" designation for an emergency exit door or emergency exit window. The labels must state "Do Not Block" using letters of at least 25 mm (1 in), in a color that contrasts with its background.

F. Test Conditions (S.5.6)

To test window retention and emergency exit release, motorcoaches and school buses must be located on a flat, horizontal surface. The temperature inside and outside the vehicle is required to be 70 to 85°F for the four hours preceding and during both tests. All windows must be installed, closed, and latched (where latches are provided), as in normal bus operation for the window retention and emergency exit release tests. In addition, seats, armrests, and interior objects must be installed for normal use and seats must be in an upright position for the emergency exit release test.

2.1.1.2 *FMVSS 220*

FMVSS 220[13] requires that school buses withstand roof crush forces encountered in rollover crashes. A force (measured in Newtons (N), and equal to 1½ times the vehicle unloaded mass in kilograms, multiplied by 9.8 m/sec^2), is applied to the school bus roof structure by means of a specified force plate. The roof crush test performance criteria require that the downward vertical movement must not exceed 130 mm (5 in).

[**] Retroreflective" material is capable of reflecting light rays directly back to the light source.

[***] Consistent with ASTM 14 "Type III" material performance criteria[14]

In addition, each vehicle emergency exit complying with FMVSS 217 must be able to operate during the full application of the force and after the force is removed. (Emergency roof exits are not required to be capable of opening during the test.). A test vehicle (i.e., test specimen) is not required to meet the emergency exit opening requirement after release of the force if it is subjected to the emergency opening requirements during the full force application.

FMVSS 220 specifies testing requirements to ensure the capability of school bus emergency exits to open as required. The exits must be open during the full application of the specified force. The exits must then open as required, after the release of all downward forces applied to the force application plate. Ambient temperature must be between $0°$ and $32°C$ (32 and $90°F$.). Vehicle doors, windows, and emergency exits must be in a fully closed position and latched but not locked.

2.1.1.3 *FMCSA*

49 CFR, Subsection 393.62 specifies emergency egress requirements applicable to all buses operated in interstate commerce.[16] Buses built after 1973 (including school buses used in interstate commerce for non-school operation) must comply with FMVSS 217 bus emergency exit requirements in effect at the time of manufacture. There are certain exceptions for buses built before September 1, 1973.

In 2005, 49 CFR, Part 393, Subsection 393.92, which contained requirements for marking emergency exit doors using 2.5-cm (1-in)-high letters and a red "marker" light, was deleted, as part of revisions for consistency with FMVSS 217.

Subsection 396.3 of CFR, Part 396 requires that push-out windows be inspected every 90 days, and that test records be kept.[17] (Note: Although Subsection 396.3 also requires that emergency doors and emergency door marking be inspected and records be kept, the requirement for such doors and markings was deleted in 2005, as indicated above.)

2.1.2 Historical Background

The following subsections summarize the development history of FMVSS 217 requirements applicable to all buses, including motorcoaches and school buses, and related requirements in FMVSS 220 (school bus rollover tests relating to emergency exit operation), as well as FMCSA, Parts 393 and 396. The focus is on buses with a GVWR of more than 10,000 lbs and only final rules are included. However, Appendix A contains a list of all Federal Register notices affecting buses, regardless of their category of service and GVWR.

2.1.2.1 ICC

Buses operated in interstate service were first subject to Interstate Commerce Commission (ICC) Bureau of Motor Carrier (BMCS) safety regulations in 1937 – a "bus" was then defined as "any motor vehicle designed and used for carrying passengers."

The ICC first issued requirements for bus emergency exits on May 15, 1952,[18] for common carrier buses seating more than eight passengers. The regulations addressed three types of glazing / window requirements: 1) construction (including unobstructed areas for means of escape), and push-out windows (if laminated glass was not used); 2) prohibition of bars that obstruct the window opening; and 3) exit marking.

The 1952 rule required that windows have sufficient area to contain an ellipse with a major axis of 18 inches and a minor axis of 13 inches. The ICC required a total exit area in square inches equal to the number of seats multiplied by 67, and that no less than 40 percent of the prescribed glazing or opening be located on one side of any bus.

The push-out emergency window exits or laminated glazing were required to be identified as such by clearly visible and legible signs, lettering, or decals. The push-out or laminated glass window marking was required to include appropriate wording to indicate that it was an escape window and the method used to exit in an emergency. In addition, emergency exit doors were required to be marked by "Emergency Door" or "Emergency Exit" with letters at least one-inch high and a red light that was required to operate at least one-half hour after sunset to one-half hour before sunrise.

The BMCS (now FMCSA) and the National Highway Safety Bureau (now NHTSA) motor carrier and vehicle regulations were transferred from the ICC on April 5, 1967, to the Department of Transportation after it was established.

2.1.2.2 NHTSA and FMCSA

On May 15, 1972, NHTSA issued a notice containing a final rule for FMVSS 217 "Bus Window Retention and Release,"[12] with an effective date of May 1, 1973. These requirements were based on the 1952 ICC motor carrier regulations for intercity buses, but also became applicable to all buses (except for school buses unless they were equipped with emergency exits), including transit buses and charter buses operated on highways, even if not operated in interstate commerce. The 1972 FMVSS 217 regulation specified: 1) new minimum retention requirements for all windows and 2) included requirements for emergency exit surface and opening size dimensions, opening force limits, and emergency exit marking and instructions for emergency exit operation.

FMVSS 217 included the new requirement that all bus (other than school bus) side and rear emergency exit push-out windows were required to be retained in a manner to assure adequate resistance to outward forces. All buses (other than school buses) were required to provide at least 67 square inches of emergency exit area per seating position, of which at least 40 percent had to be on each side to the bus. All buses, other than school buses, were also required to have push-out windows and a rear emergency door to provide unobstructed emergency exit area. If the bus construction precluded the installation of a rear emergency exit door, a roof exit was required.

The minimum unobstructed bus emergency exit opening size was increased, from the smaller original BMCS opening that could pass a 46 cm by 33 cm (18 by 13 in) ellipse, to one that could pass a 50 cm (20 in) wide by 33 cm (13 in) high ellipse. The size was increased because the then National Highway Safety Bureau (NHSB) (now NHTSA)-sponsored 1970[19] study by the University of Oklahoma Research Institute (OKRI) found that 2 out of 9 subjects who attempted to exit through a 46 by 33 cm (18 by 13 in) ellipse could not fit through it. The rule also included the option to use other than push-out windows, such as doors and panels meeting emergency exit requirements. In addition, the alternative of an emergency roof exit hatch, instead of a rear emergency exit door, was permitted to provide design flexibility for rear engine buses, while also providing for emergency egress in rollover situations. Push-out windows or other emergency exits were not required for school buses due to the risk of children falling from the windows. However, if such windows or other emergency exits were installed on school buses, they were required to meet all FMVSS 217 emergency exit requirements.

On June 10, 1972, the then BMCS issued a notice stating that revisions to FMVSS 393, including some clarifications, would become effective July 1, 1973. Buses built before September 1, 1973, were permitted to comply with the earlier 1968 regulations or FMVSS 217, at the option of the bus operator.

NHTSA issued amendments to the FMVSS 217 final rule on March 6, 1973, effective September 1, 1973, deleting the torque requirement for the emergency exit release mechanism emergency exits. Certain figures were also revised and added to make clear that exit access areas, including roof exits, must be such that the occupant has access when the bus is upright or on its side.

Since 1973, NHTSA has issued several other revisions to FMVSS 217 (and FMVSS 220).

On May 2, 1974, NHTSA issued a notice that exempted buses from FMVSS 217 emergency exit requirements that transport prisoners with an effective date of June 4, 1974. This was in response to a request from the Department of Justice.

NHTSA issued a notice on December 31, 1975, which expanded the definition of school bus in 49 CFR, Part 571, Subsection 571.3, to apply to buses operated in interstate commerce equipped to carry the driver and passengers to school related events.

NHTSA issued two notices on January 27, 1976, both with effective dates of October 27, 1976, that amended FMVSS 217 and 220 requirements. The first notice revised FMVSS 217 to include specific requirements for school bus emergency doors. Specific requirements were included, such as installing at least one rear emergency door or at least two side emergency doors. In addition, door release mechanisms were required to: 1) without the use of remote controls or tools, 2) be connected to the engine ignition to prevent exit operation while the bus was moving, and 3) sound a warning when open and release unlatched. The emergency door marking was required to state "Emergency Doors," using 2-inch (5 cm) high letters with a contrasting color; arrows, in a contrasting color to indicate which direction the release would be located, were required adjacent to the exit. The second notice contained a final FMVSS 220 rule that required school bus emergency exits to be operable, i.e., opening as specified in FMVSS 217, during and after the roof (crush) force application.

NHTSA issued a notice on June 3, 1976, that amended FMVSS 217 with an effective date of October 27, 1976. The final rule permitted the option to install a left-side door and a push-out 16 in by 48 in rear window, as school bus emergency exits. More specific emergency exit labeling on non-school buses, as well as school buses, were also required, in order to provide additional guidance regarding the location of emergency exits and the actions necessary to release and open the exits. School buses were required to use exit marking with 15 cm (6 in)-high lettering and exit instructions with letters at least .95 cm (3/8 in) high.

NHTSA issued a notice on August 26, 1976, that delayed the effective date of the school bus definition to April 1, 1977. In addition, FMVSS 220 was revised to state that roof exits covered by the roof force application plate mandatory for the test were not required to be operational while the force plate was in place.

On November 2, 1992, NHTSA issued a notice that amended FMVSS 217 to revise the title to "Bus Emergency Exits and Window Retention and Release." FMVSS 217 was also amended to improve school bus emergency exit provisions by: 1) increasing the number of school bus emergency exits; 2) considering seating capacity; and 3) revising requirements to add exit doors, roof exit hatches, and, at the option of the manufacturer, side emergency exit doors roof exits or push-out emergency windows, in that order. Requirements to increase emergency exit door and emergency roof exit conspicuity by using retroreflective marking were also included. The effective date was May 2, 1994.

NHTSA issued a notice on May 9, 1995, that amended FMVSS 217 to permit the installation of two emergency egress windows as an alternative to a single emergency exit door on school buses, permit non-school buses to meet certain school bus emergency exit requirements, permit the use of other than push-out windows, and provide additional clarity for retroreflective tape use to mark the exterior of school bus emergency exits. Manufacturers were also permitted to install emergency exit windows other than the push-out type, including vertically sliding windows, or an additional side emergency exit door on school buses. However, a mixture of sliding and push-out emergency exit windows was not permitted on school buses (with the exception of a rear push-out window), to prevent confusion as to how to open a particular emergency exit.

On April 19, 2002, NHTSA issued a notice that amended FMVSS 217 to reduce the likelihood that school bus wheelchair anchorages would be installed in locations that would block emergency egress. In addition, that rule contains a new requirement that doors and exits currently labeled as Emergency Doors or Emergency Exits be also labeled with "Do Not Block," in a color that contrasts with the background of the label. The effective date was April 21, 2003, but NHTSA later issued two notices that delayed the effective date to April 21, 2006.

NHTSA issued a notice on August 12, 2005, that amended FMVSS 217 to further respond to petitions for reconsideration related to wheelchair anchorage placement where they could block emergency exits. NHTSA agreed with the petition concerning the placement of the parallelepiped in the tangent to the opening of the rear door and revised the final rule to prohibit anchorages that are raised, flush, or recessed in the school bus beneath the beneath the parallelepiped. . NHTSA denied the petition to allow anchorages blocking access to emergency exit windows. In addition, NHTSA retained the original required "Do Not Block" label, and clarified that the label was required only for wheelchair anchorages. The effective date was revised to April 24, 2006.

FMCSA issued a notice that amended Subsection 393.62 on August 15, 2005, to make its bus exit requirements compatible with the NHTSA regulations. Buses constructed before September 1, 1973 continued to have the option of complying with the prior to 1973 required regulations or the FMVSS 217 requirements in effect when the bus was manufactured. (The 393.62 regulations had not been updated since 1972.) In addition, FMCSA deleted the longstanding requirement to identify emergency exit doors with marking using 2.5 cm (1-in)-high letters and a red marker light that was operational and visible during darkness hours.

2.1.3 Recent NHTSA Regulatory-Related Activities

2.1.3.1 *Comprehensive NHTSA Research Plan – August 2007*

The background section of the NHTSA Research Plan[3] includes motorcoach statistics for 1996-2005. While motorcoach crashes are rare, they can cause a significant number of fatalities or serious injuries. Three events in 1999, 2004, and 2005 caused a large number of fatalities, ranging from 15 to 23 persons. The Research Plan lists pertinent NTSB recommendations and describes the crashes that led to these recommendations.

The Research Plan mentioned the 2003 cooperative NHTSA Transport Canada research program[20] for glazing retention and structural integrity, which provides emergency exit-relevant information. The 2003 program study concluded that considerably more effort was needed to establish the effect on emergency egress of different glazing materials and configurations.

The Priority Strategies section of the Research Plan describes the prevention, mitigation, and evacuation strategies chosen by NHTSA to address the identified safety issues, based on the following considerations.

- Cost and duration of testing, development, and analysis required;
- Likelihood that the effort would lead to the desired and successful conclusion;
- Target population and possible benefits that might be realized;
- Anticipated cost of implementing the ensuing requirements into the motorcoach fleet; and
- [NTSB] "Most Wanted List."

The three topic areas that NHTSA selected for research include:

1) emergency egress, e.g., the number and type of emergency exits;
2) Signage, including the use of photoluminescent materials; and
3) Illumination, e.g., emergency lighting.

Specific research approaches are described for each topic area.

2.1.3.2 *FMVSS 217 Regulatory Review Assessment*

This June 2007 paper,[21] prepared by NHTSA, provides a regulatory review and assessment of FMVSS 217. Safety problems pertaining to window retention and release, and bus emergency exits, such as occupant ejection or problems in exiting the bus, e.g., entrapment, jammed windows, are reviewed.

Several databases were reviewed containing bus occupant casualty data. Statistics were presented relating to type of bus involved, occupant age, and occupant ejection and extrication casualties.

The major part of the review and assessment consists of a technologies section derived in its entirety from a 2005 engineering assessment reviewed in the following section.

2.1.3.3 *Battelle Engineering Assessment*

In 2005, NHTSA contracted with Battelle to provide support for its regulatory review plan.[22] The Battelle report reviewed and assessed FMVSS 217 and FMVSS 220. The purpose of the assessment was to identify new technologies for illuminating and labeling bus exits. Battelle conducted a literature search and contacted bus and product manufacturers. Emergency exit operation and glazing materials for both motorcoaches and school buses were extensively discussed. Battelle compared existing FMVSS 217 bus and school bus marking requirements and lighting technology standards and identified new emergency exit lighting technologies.

Volpe Center staff completed a technical review of the Battelle assessment.

2.1.4 University of Oklahoma Research Institute Studies

NHTSA (then the NHSB)) contracted with the University of Oklahoma Research Institute (OKRI) to conduct two research studies, which were published in 1970 and 1972 pertaining to motor vehicle post-crash "escapeworthiness," which were published in 1970 and 1972.[19,23] Intercity motorcoaches and school buses were included in the scope of work. The two studies, as well as a third study completed by OKRI in 1978 that focused only on motorcoach emergency evacuation,[24] and provide the only known previous U.S. bus evacuation experimental data, prior to the Volpe Center experiments described in Chapter 3. While the three reports are summarized below, specific sections of additional interest in the 1970, 1972, and 1978 reports are discussed in Chapters 4-9.

2.1.4.1 *1970 Motor Vehicle Escapeworthiness Research Study*

This OKRI study was initiated to develop information that could be used as a basis for establishing minimum vehicle design standards to reduce motor vehicle post-crash injuries.[19] Accordingly, the study examined various post-crash factors affecting survival in crashes and other emergency situations involving passenger cars, school buses, and intercity motorcoaches. The bus research task involved the conduct of experiments to determine emergency egress rates from intercity and school buses. (Note: Separate window retention measurements were also conducted.)

The major independent variables for the OKRI bus experiments included the vehicle (type, size, location, and number of exits), passengers (age, gender, weight and height, and total number), and the environment (day or darkness conditions). Dependent variables included time to escape, occupant behavior, and injuries.

Darkness during selected trials was simulated by the use of special goggles.

While a series of school bus egress experiment trials was conducted using various exits, including the side windows and rear side doors, and roof hatches, the remainder of this summary provides information applicable only to the school bus roof hatch experiment trials, but all intercity bus egress experiment trials.

A. School Bus

Sixty subjects representing all twelve grades from the university laboratory school, with approximately half boys and half girls, participated in the school bus experiment trials (with the exception of a few tests where one or two subjects were treated for minor injuries). The main variable was body breadth (shoulder, elbow to elbow, and hip, sitting) in relation to the size of the exit. Other characteristics such as age, etc., were also considered to provide a reasonable cross section.

Depending on the type of trial, school bus escape routes included the front door, and 6 push open emergency exit windows (each 62 by 51 cm (24.5 by 20 in) high), and a side rear emergency exit door 70 cm wide by 27 cm high by 50 by (25 7/8 in).

An overturned school bus mock-up was used for certain escape trials with one 61 by 102 cm (24 by 40 in) and two 61 by 61 cm (24 by 24 in) size emergency roof exit hatch openings. A platform using layers of mattress was positioned under the roof hatch opening exits.

Two series of school bus trials using a rear emergency exit door and emergency roof exit hatch mockups as the escape route, while the bus was on its side, were used.

- Series 1
  - Trials 1-3 used the rear door and different sizes and locations of the roof hatch.
  - Trial 4 used two 61 by 102-cm (24 by 40-in) roof hatches. The subjects wore goggles to simulate darkness.
- Series 2 of the roof hatch trials repeated Trials 1-4, except that the subjects did not wear goggles to simulate darkness during Trials 1-3.

B. Intercity Bus

For the intercity bus experiments, suitable subjects were located through use of newspaper advertising and direct contact. 15 subjects were selected to represent a distribution above and below the 95th percentile and a reasonable age range. All subjects were paid and advised of the possible hazards and screened for latent injuries such as back or cardiac conditions.

A *GMC Model PD4104* bus with a seating capacity of 39 passengers and equipped with a lavatory was used.

Egress by 10 subjects was also conducted using three emergency exit window opening sizes were simulated by major and minor ellipses of:

- 45 and 33 cm (17 ¾ and 13 in),
- 51 and 33 cm (20 and 13 in), and
- 61 and 43 cm (24 and 17 in).

Table 2-1 shows the egress rate results for the three window ellipsoid sizes (which are further discussed in Chapter 5.

Table 2-1. 1970 OKRI Subject Egress Rates for Escape Through Various Size Ellipses

WINDOW TRIAL	ELLIPSE DIMENSIONS cm (in)	NUMBER OF SUBJECTS WHO ESCAPED	AVERAGE TIME PER SUBJECT (sec)	EGRESS RATE (ppm)
1	45 and 33 (17 ¾ x 13)	7 [#]	4.31	13.9
2	51 and 33 (20 x 13)	9 [# ##]	4.26	14.1
3	61 and 43 (24 x 17)	10	2.46	24.4

[#] Subject tried but could not escape through exit.
[##] Subject injured during Trial 2 and could not participate

Additional upright escape route trials included use of:

- Rear emergency exit door and the side emergency exit windows;
- Front door, rear emergency exit door, and side emergency exit windows;
- Emergency exit windows only on the left side and the rear emergency exit door; Replication of the first trial to explore learning effects; and
- Side emergency exit windows, rear emergency exit door and special emergency exit door in left side of the bus.

An additional four trials were conducted to repeat the first four trials listed above with the subjects wearing goggles to simulate darkness. Trial data was not reported in the 1970 report, except for the emergency exit window size (see above and Chapter 5)

C. Findings

The 1970 major findings included:

- Escape from buses is significantly affected by:
    - Post-crash bus position,
    - Exit size,
    - Location of exit in relation to its height above the ground,
    - Weight of the emergency exit windows or doors in some positions, and
    - Darkness;
- Physical size and age significantly affect escape time from buses;
- Push-out windows pose significant problems as an escape route from a bus, being too difficult to open;
- Use of rear exit doors and roof hatches significantly reduce escape times; and
- Of three window sizes investigated, the largest (61 cm-wide by 43-cm high (24 by 17 in)) resulted in significantly shorter escape times.

2.1.4.2 *1972 Motor Vehicle Escapeworthiness Research Study*

This OKRI study was a follow-up to the 1970 OKRI study and further examined various bus survival factors affecting passenger during post-crashes and other emergency situations involving passenger cars, school buses, and intercity motorcoaches.[23] The study results provided additional important information to NHTSA relating to the requirements that were included in FMVSS 217. The bus research task involved a series of emergency egress rates from school buses and intercity buses. (Separate window retention measurements were also conducted.)

The major independent and dependent variables for the 1972 experiments conducted by OKRI were the same as in the 1970 study. Due to the variations in school bus and intercity bus emergency exit window design and operation, the remainder of this summary description of the 1972 study is limited to the intercity bus experiment trials.

An upright 39-passenger *MCI Model PD4204* bus with a lavatory was used during the intercity bus experiments. A platform using layers of mattress was positioned under the emergency exit windows and rear emergency exit doors. All emergency exit window escapes were conducted with the window tied back to avoid injury from a window falling shut during and after escape.

The 38 subjects who participated were selected based on a survey of passengers observed at the local Oklahoma City bus station and confirmed by contact from the bus company. The only major difference in the selected group versus the observed group was the absence of females over 60 years of age. All subjects received payment.

The experiment escape trials included use of various exit routes: Darkness during selected trials was simulated by the use of special goggles.

As Table 2-2 shows, the intercity bus experiment trials resulted in high egress rates because several escape paths were used at the same time and the distances to the ground were shorter. The egress rate is expressed in persons per minute (ppm).

Table 2-2. Summary of 1972 OKRI Intercity Bus Egress Rates

TRIAL		AVERAGE EGRESS TIME PER PERSON (sec)	RANGE (+/- sec)	TOTAL EGRESS TIME FOR 38 PERSONS (sec)	EGRESS RATE (ppm)
Number	Condition				
1	Emergency window exits only With goggles	2.21	n/a	84.1	27
2	Emergency exit windows & left rear side emergency exit door No goggles	0.97	n/a	37.0	62
3	Same as 2 With goggles	0.83	n/a	31.6	72
4	Emergency exit windows, left rear side emergency exit door & front door No goggles	0.57	n/a	21.8	105

The 1972 report stated that the lavatory partially blocked the rearmost side window exit, making it difficult to push window open. Some male subjects voluntarily held the window exits open for other to escape; otherwise, the escape times would have been considerably longer.

The principal study findings include the following:

- A maximum acceptable time for passengers to escape from a bus using only half of the exits should not exceed 90 seconds.

- Hazards exist in the form of sharp objects, which would cause injury to passengers attempting to exit through a window.

- The new (at that time proposed) bus exit safety standard should be more strict, and similar standards should be provided for school buses.
- Based on strength measurements for females and children, more attention should be given to operating methods and force requirements, as well as the interpretability of operating instructions.

2.1.4.3 1978 Intercity Bus Evacuation Study

This OKRI study focused only on intercity buses.[24] The objectives were to:

- Determine typical circumstances of intercity bus crashes and other emergency situations and important variables affecting evacuation.
- Determine a profile of a typical intercity bus passenger load including such variables as height, weight, age, and gender.
- Develop several scenarios representative of worst-case conditions.
- Conduct empirical tests of evacuation performance for the selected conditions.

A bus passenger profile was developed using observations of 959 persons at the Oklahoma City bus terminal in January 1977. Variables selected were: age, gender, weight, height, and clothing type.

The 1978-dated study (conducted in 1977) used the same type of independent and dependent experiment variables as used in the 1970 and 1972 studies.[19][23]

The 135 subjects who participated were well matched to the age and gender distribution of the intercity-bus-riding population. For other parameters, there was some variability among each of the subject groups of 45 persons.

The *GMS Model PD4207* bus seating capacity was 45 passengers (no lavatory was provided). The bus was equipped with eight push-out side windows that met FMVSS 217 emergency exit window requirements, in effect in 1977. The height of the windowsill was 1.8 m (6 ft) above ground level when the bus was upright. A pop-out roof hatch 54.6 cm by 49 cm (21.5 in by 19 ¼ in) was located in the rear half of the bus roof. The front door opening, when available, for egress was 71 cm wide by 2 m in height (28 in by 7 ft).

The worst-case evacuation scenario for the upright bus selected was the front door blocked under darkness conditions (Trial 1). The worst-case scenario was a bus on its side with the front door blocked (Trial 3). Several subjects simulated injuries in each trial; during certain trials, subjects were slightly injured during the egress process.

Egress rate data for the six experimental conditions are shown in Table 2-3. The report stated that there was a great disparity in the total time that each window was used to escape and the number of persons escaping through each of the different exits. Different proportions of the 45 subjects chose different egress paths in each trial. The highest values in the range column are associated with either the subjects who had to open an exit or those who were pretending to be injured.

Table 2-3. Summary of 1978 OKRI Intercity Bus Egress Rates

EXPERIMENT TRIAL		AVERAGE EGRESS TIME PER PERSON (sec)	RANGE (+/- sec)	TOTAL EGRESS TIME (sec)	EGRESS RATE (ppm)
Number	Condition				
1	Upright bus 8 side windows used Front door not used Darkness	2.4	3.7	108.54	25
2	Upright bus 6 side windows & Front door used # Darkness	1.7	.04-44.1	77.63	35
3	Overturned Bus on Right side 4 side windows & Roof hatch used ## Darkness	3.2	1.8-24	142.8	19
4	Overturned Bus on Right side 3 side windows & Roof exit hatch used ### Darkness	2.2	1.2-20	98.54	27
5	Overturned Bus on Right side 3 side windows, & Roof hatch used### With Emergency Illumination	2.5	1-26.5	112.67	24
6	Same as Trial 5, With windshield already open	1.3	0.9-18.0	56.04	48

\#   Subjects were not told that the front door would be available.
\#\#   Subjects broke thru and also d through front windshield
\#\#\#   The roof hatch and front windshield remained open from Trial 3 and were used

The average time to open an emergency exit window in the simulated-darkness conditions, with the bus upright was 14.6 sec (range: +/- 9.7 sec). In darkness and with the bus overturned, the average time to open a window increased to 18.6 sec (range: +/- 14.1 sec). Egress through the windows was more difficult when the bus was overturned.

All of the egress times were longer than during the 1972 experiments because the front door was blocked during certain trials and subjects had to climb through other exits, such as side windows, roof hatch, front windshield, in order to exit from the bus.

The 1978 study major findings include the following:

- A standard for maximum evacuation time should be considered.

- Some type of ladder or "toehold" on the inside and outside of the bus to improve their use by passengers and at least three roof hatches should be provided.

- Clear instructions should be provided at all exits for passenger use. (The *Human Engineering Guide to Equipment Design*[25] was cited.) A type of escape instruction such as is used on aircraft should also be provided to passengers.

- An emergency illumination system should be considered for buses. This system should be able to function after a crash to provide illumination and reduce evacuation time, as well as assist in the first-aid treatment of passengers.

- Window hinges used on buses should have a performance requirement that would prevent the window from breaking off under the loads expected from passengers pushing the windows open rapidly for escape and when passengers attempt to hold onto the window to lower themselves to the ground from the side of an overturned bus.

2.1.5 FMCSA Pre-Trip Safety Briefing Guidance

As a result of a special investigation of selective motorcoach issues,[8] NTSB issued two recommendations:

- Provide guidance for the minimum safety information to be included in briefing materials for motorcoach operators (HS-99-7); and

- Require motorcoach operators to provide passengers with pre-trip information (H-99-8).

FMSCA issued a notice on August 28, 2007, containing that guidance to motorcoach operators.[26] The minimum safety topics for the motorcoach passenger safety awareness plan relating to emergency evacuation provided are: driver direction, emergency contact, location and operation of push-out windows, roof vent, and side doors, including the emergency door release located on the dashboard, and restroom emergency push signal switch. Alternative means permitted to provide this information include: informational pamphlets distributed to passengers during boarding; or the use of several other methods after passenger have boarded. These methods include informational pamphlets located in seat back pouches; an oral presentation by the driver, similar to that provided by airline flight attendants; or automated audio or video presentations broadcast over the motorcoach audio or video system. The recommended timing and frequency were: 1) before charter or tour buses move and 2) before buses are moved at major terminals used for fixed motorcoach routes. FMCSA developed "model" brochures for use by motorcoach companies to provide safety information to their passengers. These brochures (see Appendix D) address the safety topics specified in the FMCSA pre-trip guidance passenger safety awareness plan and include illustrations and instructions for operating pull and lift emergency window exits and roof hatches.[27][28]

2.2 OTHER U.S. TRANSPORTATION VEHICLE REQUIREMENTS

Federal Aviation Administration (FAA), Federal Railroad Administration (FRA), and the United States Coast Guard (USCG) each have requirements for passenger vehicle emergency egress.

In addition, the American Public Transportation Association (APTA) has published three industry standards for passenger rail car emergency systems. These three standards represent a systems approach to identifying, reaching, and operating emergency exits, as well as identification and operation of rescue access locations.

Appendix C contains a series of tables that provide a comparison of NHTSA bus and school bus emergency egress-related regulations and U.S. regulations for other transportation vehicles. These regulations, as well and the APTA standards, are cited in Chapters 4-9, where appropriate.

2.2.1 FAA

FAA requires that passenger aircraft manufacturers and airline operators comply with the following.[29]

- Demonstrate passenger evacuation from an aircraft in 90 seconds.
- Meet extensive specific performance requirements for the minimum number, type, and location of emergency exits, and other emergency evacuation components. The number, type, and size of exits required are based on the number of passenger seats.
- Illuminate emergency exit signs by an independently powered electrical source; emergency lighting is also required to be independently powered.
- Provide "floor proximity emergency exit path marking."

2.2.2 FRA

FRA requires that passenger rail operators and manufacturers comply with extensive emergency exit regulations contained in 49 CFR, Part 238 and 239.[30] (FRA recently issued a revision to Part 238; some requirements are effective immediately for all equipment and some are effective only for new equipment.)[31]

Passenger rail cars must comply with the following:

- Meet extensive specific performance requirements for the minimum number, type, and location of emergency exits, rescue access locations, and other emergency egress components. Certain requirements are applicable only to new equipment.

- All doors that could be used for passenger emergency egress, or rescue access by emergency responders, must be conspicuously marked on or adjacent to each exit / access location. These marking must be luminescent material on the interior and retroreflective material on the exterior[****]. Instructions for operating the doors for emergency egress, emergency exit windows, and rescue access locations must also be marked on or adjacent to each exit / access location

- Emergency lighting is required for each new passenger car with minimum illumination levels measured at the floor adjacent to the car doors, as well as along the aisle, at armrest level. Roof access points are required on all new equipment.

- Passenger information must be provided that explains the operation of the emergency exits; this may take the form of passenger instruction cards in seat backs or other means.

2.2.3 APTA PRESS Standards

American Public Transportation Association (APTA) passenger rail equipment safety standards (PRESS) contain specific performance criteria for emergency lighting,[32] emergency signage,[33] and low-location exit path marking.[34] The requirements contained in each of the three standards vary according to whether the passenger rail equipment is existing or new construction.

PRESS standards were developed to assist passenger railroad operators in complying with FRA regulations.

FRA has indicated that it plans to incorporate the three APTA standards by reference in the next revision to the Part 238 passenger rail car safety standard regulations.[31]

2.2.4 USCG

USCG requires that all passenger vessels provide at least two means of escape from all passenger-accessible areas.[35]

Additional USCG requirements specify performance criteria for emergency exit marking, emergency lighting, as well as the vertical travel distance to exits, stairway sizing, and other provisions relating to emergency egress.

[****] More information related to conspicuity, as well as luminescent and retroreflective material properties, is discussed in Chapter 8.

For higher passenger density dinner excursion and gambling vessels, Navigation and Vessel Inspection Circular (NVIC) 8-93 provides a description of equivalent alternatives for meeting USCG requirements for means of escape, main vertical zones, and safe refuge areas.[36]

In addition, USCG also recognizes and requires that U.S. passenger ships that operate in international waters comply with International Maritime Organization / Safety of Life at Sea (IMO / SOLAS) requirements, including those related to emergency egress.[37]

2.3 INTERNATIONAL BUS REGULATIONS

2.3.1 Economic Commission for Europe (ECE)

The ECE has issued two "uniform provisions" for buses: ECE 36 (construction of public service vehicles)[38] and ECE 107 (general construction for double deck buses),[39] contain requirements pertaining to bus emergency egress. Section 5.5.6.6 of ECE 36 states "There shall be at least two internal lighting circuits such that failure of one will not affect the other." A circuit serving only permanent entry and exit lighting can be considered as one of these circuits." In addition, Section 5.5.5 specifies that emergency lighting circuits are exempt from inclusion with all of the other electrical circuits that could be disconnected by operation of the emergency switch (a device to disconnect the main battery in the event of an electrical fire).

2.3.2 Canada

The Canadian Transport agency issued a bus safety review report in 1998.[40] The review covered both school buses and buses other than school buses. The report stated that the Canadian bus regulations are in harmony with FMVSS 217, except that bilingual exit marking is required and only push-out windows are permitted.

2.3.3 Australia

All Australian motorcoaches must comply with the Australian Design Rules (ADRs), in addition to the ECE requirements. The ADRs specify design standards for vehicle safety and emissions. ADRs pertaining to motorcoaches are ADR 44/02 "Specific Purpose Vehicle Requirements"[41] and ADR 58/00 "Requirements for Omnibuses Designed for Hire and Reward" for buses for public transport buses.[42] ADR 44/00 was first established in 1986 and applies to vehicles built from July 1, 1988. ADR 44/02 applies only to large buses with more than 16 passengers including the driver and crew. Revisions have been made since 1986 with 44/01 applying to vehicles built from July 1, 1992 and ADR 44/02 applying to vehicles built from July 1, 1993.

Private-use buses must comply with ADR 44/02, but public-use buses (99 percent) of the fleet may be built to comply with either 44/02 or 58/00. Even though 44/02 imposes more requirements for emergency egress than 58/00, most manufacturers elect to comply with ADR 44/02 because it allows greater design flexibility.

Australia became a signatory to the United Nations / ECE agreement in 2000 and is in the process of harmonizing its vehicle regulations with those of the ECE. That process is nearly complete, and is apparently in effect for heavy (i.e., large) buses, pending confirmation from Australian government). However, Australia reserves the right to impose additional requirements where it finds a public safety benefit sufficient to justify the cost.

ADR 44/02 includes the following emergency egress elements that are significantly more stringent than the ECE regulations:

- Emergency doors and windows with a bottom edge more than 1,000 mm (39 in) above the ground shall have a means to assist occupants in descending to the ground, such as footholds, no more than 500 mm (20 in) apart. Emergency doors and windows with a bottom edge more than 2,000 mm (79 in) above the ground shall be equipped with self-supporting steps or equivalent to assist occupants in descending to the ground.

- Interior "exit" signs shall be permanently illuminated while the vehicle is in operation, and shall be illuminated or self-illuminating (e.g., photoluminescent) for at least 15 minutes after the vehicle ceases operation or after loss of battery power.

2.3.4 United Kingdom

The United Kingdom (UK) regulations are contained in "Public Service Vehicles (Conditions of Fitness, Equipment, Use and Certification) Regulations (No. 257)" apply only to public service vehicles designed to carry more than 8 passengers, for hire, such as buses and charter coaches.[43]

Contacts with the UK Department of Transport indicate that efforts to harmonize the British standards with the ECE regulations are underway.

2.4 OTHER RELEVANT INFORMATION

Other information identified during the literature search considered relevant is summarized below.

2.4.1 NTSB Reports

In addition to the NTSB accident investigation reports [7,10] and the two special investigation reports [8,9] cited earlier relating to motorcoaches, numerous other NTSB accident reports for both motorcoaches and school buses were reviewed. These reports provided important information relating to specific difficulties that passengers encountered during emergency egress from buses.

2.4.2 1983 NBS Technical Note 1180

In 1983, the National Bureau of Standards (NBS) published a study for the U.S. Department of Labor that investigated the size of letters required for building emergency exit signs.[44] Volpe Center staff adapted the findings of the NBS report to develop minimum performance criteria for passenger rail car emergency exit signs. The NBS report is discussed in Chapter 9.

2.4.3 Human Engineering Guide to Equipment Design

The *Human Engineering Guide to Equipment Design*[25] contains general human factors engineering criteria for equipment design. A labeling section provides extensive information useful for considering emergency exit signage design.

2.4.4 MIL-STD 1472F Design Criteria Standard for Human Engineering

MIL STD 1472F Design Criteria Standard for Human Engineering[45] contains general human factors engineering criteria for military systems, subsystems, equipment, and facilities. A labeling chapter provides extensive information useful for considering emergency exit signage design.

2.4.5 UK Strength Data – 2002

Nottingham University conducted research on behalf of the Consumer Affairs Directorate of the Department of Trade and Industry (DTI), United Kingdom (UK) in 2002. Two reports are available, which provide designers with ergonomics data for use in the design of safer products.[46][47] During Phase 1, data was gathered for six different force exertions. During Phase 2, data were collected for eight strength measurements. Data were collected from about 150 British citizens, 2 to 90 years of age, and balanced across age groups and genders.

2.4.6 TCRP 100 Manual

The Transit Cooperative Research Program *(TCRP) 100 Manual*[48] provides a logical and consistent set of methods and techniques for evaluating public transportation vehicle and facility passenger capacities. Part 4 of the Manual contains a calculation method for boarding and deboarding passengers, as well as empirical data for boarding and deboarding rates for various types of vehicles.

2.4.7 Volpe Center Alaska Bus Egress Study

Volpe Center staff previously conducted a study for the U.S. Forest Service to determine physical requirements for transportation facilities near the Mendenhall Glacier in Alaska.[49] In order to estimate the number of bus loading bays needed at various locations, it was necessary to learn how much time is needed per passenger for loading and unloading. Because no data were available applicable to tourists in a cold climate, the authors elected to conduct the measurements themselves. The raw data were not included in the published report, but were provided to the authors of this study.

2.5 OTHER

Additional reports or documents of interest are cited where appropriate in Chapters 4-9.

2.6 SUMMARY

The literature review was valuable in locating a wide variety of information. In particular, the previous NHTSA-sponsored research studies provide the only known previous U.S. bus evacuation experimental data, prior to the Volpe Center experiments described in Chapter 3. As such, they provided extensive valuable background useful in developing the Volpe Center motorcoach egress human factors experimental plan.

The review of U.S. regulations and industry standards applicable to other transportation vehicles, and of international standards, and other relevant research provided an extensive useful resource during the preparation of Chapters 4-9.

3. MOTORCOACH EGRESS HUMAN FACTORS EXPERIMENTS

Volpe Center staff designed and conducted a series of field observations and controlled experiments to learn more about the human factors aspects relating to motorcoach emergency egress. This chapter summarizes the type of activities conducted, including:

- Naturalistic observations of passengers exiting from motorcoaches located at a large bus terminal under normal conditions;

- Development of instrumentation to measure opening forces, primarily for emergency exit windows and roof emergency roof exit hatches;

- Field visits to inspect current design motorcoaches, in terms of front door, emergency exit window, and emergency roof exit hatch design and marking, and conduct of force measurements for the latter two types of exits;

- Field visit to MGA test facility to use emergency exit windows and emergency roof exit hatches to exit two different models of motorcoaches; and

- Development and conduct of egress experiments using: 1) a full load of subjects for front door egress, and 2) a smaller number of persons for emergency exit window and wheelchair-access door egress, to determine egress flow rates for the three types of exits.

Specific results for the different activities are also discussed, as appropriate, in Chapters 4-9.

3.1 NATURALISTIC PASSENGER OBSERVATIONS

Normal egress time from large passenger vehicles, whether planes, trains or motorcoaches, tends to be highly variable. The most important determinants of egress flow rates (passengers per minute) for normal vehicles are the agility of each passenger in the stream and the extent to which various passengers are burdened with luggage, parcels, and small children. Distractions such as searches for misplaced items, cell-phone calls, and repacking often cause gaps in the flow. Within the range of typical passenger behavior and agility, egress flow rates vary by a factor of at least four. For an atypical distribution, e.g., a busload of nursing home residents, the egress flow rate can drop to a fraction of normal values.

Because of this variability, normal egress times have never been used as measures of the performance of the egress characteristics of vehicles. Where these characteristics are considered critical to overall design safety, as in commercial aircraft, they are evaluated in controlled experiments in which the variance in egress time is much smaller than in normal egress. The subjects are all unencumbered, able-bodied adults, who are being directed by flight attendants at each aircraft exit. Flight attendants are taught to behave as "drill sergeants" and to push any passenger who hesitates at the top of the slide. These evacuation trials can produce reasonably consistent estimates of egress time, but they are understood as "best case" values. They are typically multiplied by a safety factor of two or more to arrive at fire-safety design requirements.

Controlled emergency evacuation trials have historically excluded individuals with impaired mobility, infants, and toddlers because they introduce so much variability into the results and the risk of injury to such persons would be too great. The only readily available means of understanding egress-flow rates for such individuals is naturalistic observation.

3.1.1 Methodology

3.1.1.1 *South Station, Boston, MA*

The preferred location for observation and measurement of normal egress times and flow rates is a busy motorcoach terminal, i.e., one where all of the passengers exit. The South Station bus terminal in Boston was selected for conducting observations. The bus terminal has 25 bays surrounding a large, glass-enclosed waiting area. Several motorcoaches arrive during each hour of the day. Most of the bays can be easily observed, unless the area is too crowded, in which case ample opportunities exist to stand near one of the glass walls with a good view of passengers stepping off a bus.

Measurement of egress time and flow rate requires only a stopwatch, tally counter, and a notebook to record the data. Counting and timing begin when the bus door is opened and stop when the flow of passengers appears to have ended. Sometimes a few stragglers appear after data collection has ended. They can be simply ignored, because their behavior – searching for a lost article, repacking a suitcase, etc. – is not a concern during a catastrophic emergency evacuation.

Along with time and count, any occurrences that impede flow were entered as comments. Such occurrences included individuals with mobility impairments, persons carrying infants or small children, and travelers with bulky carry-on parcels.

Measurements were conducted at South Station on December 27, 2007, and on May 23, 2008.

3.1.1.2 *Volpe Center Mendenhall Glacier Visitors Center Study*

Observational measurements were conducted of passengers boarding and deboarding tourist buses in Alaska by Volpe Center staff for the U.S. Forest Service.[49]

3.1.2 Results

The data from the two sets of bus terminal observations at South Station, Boston, MA and the data obtained for the Alaska study are discussed in Chapter 4.

3.2 EXIT OPENING AND RELEASE FORCE MEASUREMENTS

The purpose of the Volpe Center tests was to gather precise measurements of the forces required to release and to open the emergency exit windows and emergency roof exit hatches of a representative sample of motorcoaches made available by bus operators for each of the three major manufacturers of buses operated in the United States: *MCI, Prevost (Volvo),* and *Van Hool.*

3.2.1 Existing FMVSS 217 Requirements

A critical aspect of emergency exit design is the force requirement for opening an exit. FMVSS 217 requires limits for maximum allowable force for motorcoach emergency exit windows and emergency roof exit hatches of:

- 20 lbf (89 N) in low-force areas, and
- 60 lbf (267 N) in high force areas.

Low-force areas are those close to the floor or ceiling, high-force areas are those near the shoulder height of a seated passenger — generally between 61 and 132 cm (24 and 52 in) above the floor.

Subsection 396.3 of the FMCSA regulations requires that carriers inspect all emergency exits at least once every 90 days.[11] Such inspections are conducted by opening each exit and determining that the forces required to do so are "normal."

Although NHTSA has established a test procedure that specifies how opening-force measurements must be performed[50] for its use in conducting emergency exit compliance testing, motorcoach operators do not generally use the procedure for periodic inspections of their in-service fleets.

3.2.2 Field Measurements

Volpe Center staff measured exit release and opening forces of motorcoaches that have been in revenue service, at three bus company maintenance facilities located in Massachusetts, and during the Volpe Center staff visit to the MGA test facility (see Section 3.3).

3.2.2.1 Sample Size and Sources

Volpe Center staff members visited the Peter Pan intercity and charter motorcoach maintenance garage in Chelsea, MA to take emergency exit window and emergency roof exit hatch release and opening force measurements on January 29 and March 17, 2008. The motorcoach fleet for that operator consists primarily of *MCI* buses, including both the "D" and "J" models. See websites:

   http://www.mcicoach.com/NewCoaches/PassengerCoaches/passengerDseries.htm

   http://www.mcicoach.com/NewCoaches/PassengerCoaches/passengerJ4500.htm

For *Prevost* and *Van Hool* motorcoaches, emergency exit window and emergency roof exit hatch release and opening force measurements were conducted at the *Wilson* and *Ritchie* bus company operator maintenance facilities located at East Templeton, MA and Northborough, MA, respectively, on February 19, 2008. See the following websites:

   http://www.prevostcar.com/cgi-bin/pages.cgi?page=passenger

   http://www.vanhool.be/products_car.asp?ParentID=8

At each garage, different motorcoaches were tested.

Several emergency exit window release and opening force measurements were also conducted by Volpe Center staff on the two motorcoaches used for the NHTSA-contracted roof-crush tests at the MGA test facility. The measurements were completed before the buses were damaged in a way that possibly distorted their frames and changed opening force measurements. These emergency exit window measurements were conducted on February 25, 2008.

3.2.2.2 Force Measurement Procedures

FMVSS 217 requires two separate applications of force to open an emergency exit — either as two separate release mechanisms or as a single release mechanism with applications required in two different directions. In the latter case, the initial release is typically applied through a person's fingertip(s), as shown in Figure 3-1, for an emergency roof exit hatch.

A. Release Force Measurements

Measurements for forces applied through fingertips to roof-hatches were measured with a subminiature (button-sized) load cell (*Omega Engineering LCMKD-100N$^{R)}$*) placed between a finger tip and the surface to which force was applied.

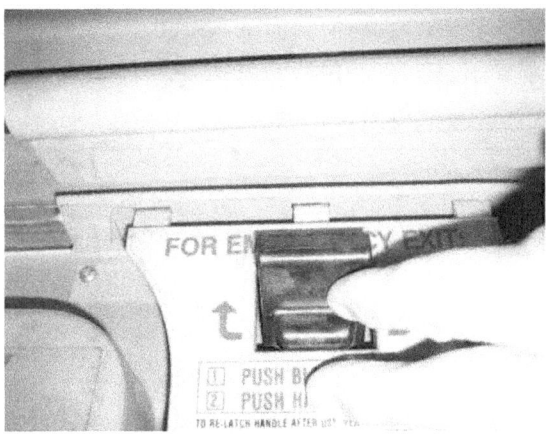

Figure 3-1.   Fingertip Release – Emergency Roof Exit Hatch

All measurements of window–release forces were made with an Omega Engineering LCCA-200 "S"-beam load cell fitted with a hook on one side and a handle on the other, as illustrated in Figure 3-2a.

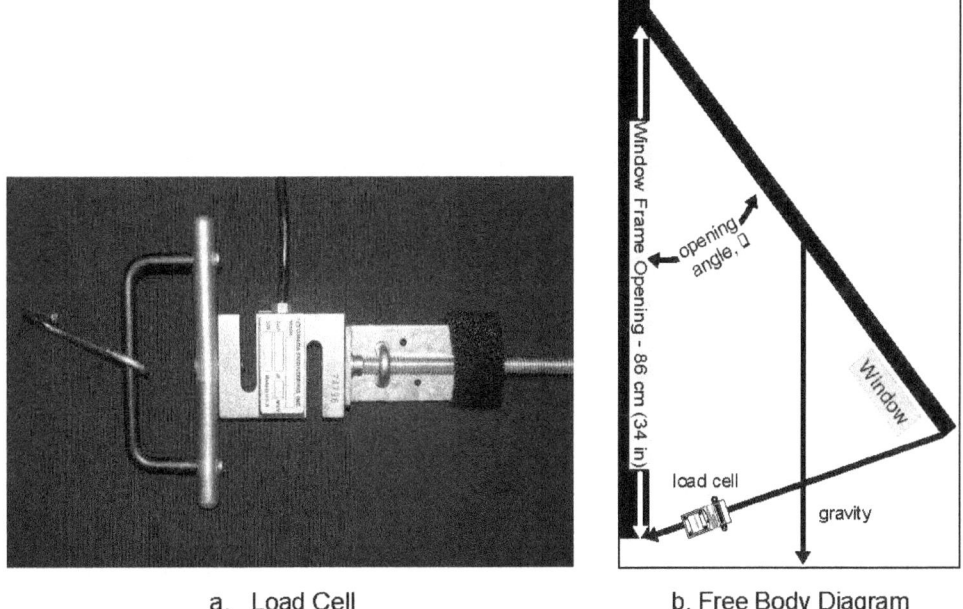

a. Load Cell   b. Free Body Diagram

Figure 3-2.   Load Cell and Free-Body Diagram of Window-Opening Forces

Excitation, amplification, and analog-to-digital conversion were accomplished using a *Dataq Instruments DI-700*^{R} data-acquisition system and the manufacturer's software. Readings were recorded at the rate of 128 samples per second, to a laptop computer. The software was configured to generate a plot of the applied force, expressed in Newtons (N), against time. Two or more release-force measurements were made in quick succession, resulting in graphic output as shown in Figure 3-3. The value where the cursor (vertical line) intersects the plotted data is reported at the left — 122.9 N in this example.

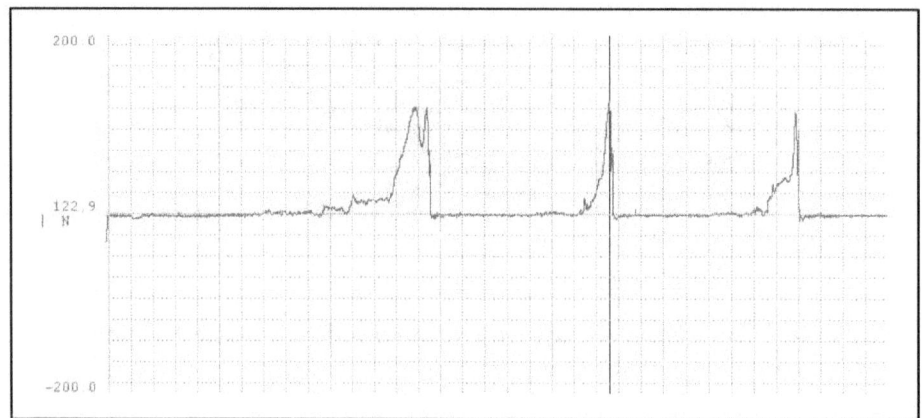

Figure 3-3. Force Measurement Graph – Peak Release Force of 122.9 N

A digital photograph was made of each emergency exit window release mechanism measured, as illustrated in Chapter 6.

Other details and descriptors of these measurements were recorded using a spreadsheet form, based on the FMVSS 217 laboratory test procedure.[50] This form was supplemented with two columns that contain the $WinDaq^R$ data file name for each measurement and the file name for the photograph of each object measured. Additional comments, such as an assessment of a window's potential for causing injury when it swings closed as an occupant exits, were recorded in the "comments" field of the data entry form.

B. Opening Force Measurements

Opening forces for roof hatches were measured by pushing them upward at the rear edge with the force applied through a load cell and recorded as described above.

To measure the forces required to open the top-hinged emergency exit windows, Volpe Center staff used the same S-beam load cell (shown in Figure 3-2a) as used for the release-force tests, since the load cell is designed to work in compression, as well as tension. Once the retention mechanism had been released, all of the windows tested opened easily. All of the weight of the window is initially carried by the hinge once it has been pushed clear of the frame. As the opening angle increases, the force that must be applied to hold it against the force of gravity increases in proportion to the sine of the opening angle. This force was measured by the load cell positioned as shown in Figure 3-2b. Before proceeding to the empirical data, a discussion of the method used to estimate the opening forces is presented to explain the forces at work.

Since the mass of the emergency exit window is distributed symmetrically, the force required to hold it at any given angle can be estimated with good accuracy by assuming that the center of gravity is located halfway between the hinge and the lower edge of the sash. Thus, the window can be modeled as a lever with 2:1 mechanical advantage, i.e., sufficient force must be applied to the lower edge of the sash to counteract one-half of the gravitational force acting on the window. The calculation formula for the force that must be applied perpendicular to the lower edge of the sash is simply:

$$F = M \times A/2 \times \sin \theta \tag{1}$$

Where:	F = Force required (Newtons)

M = Mass of window (kilograms)

A = Acceleration of gravity (9.8 m/sec^2)

θ = Angle of opening

If this opening force were applied to the emergency exit window by an individual standing on a ladder outside the motorcoach, it could be perpendicular to the window at any opening angle. However, a passenger inside the motorcoach cannot maintain a perpendicular application at the lower edge as the opening angle becomes large. A portion of the person's effort is expended pushing outward, which increases the total force. Assuming that this force, denoted F', would probably be applied in a line from the sill to the lower edge of the sash (as in inserting a prop in that that position), it can be calculated as:

$$F' = F / \cos(\theta / 2) \tag{2}$$

Figure 3-4 shows that amount of force that must be supplied to a prop inserted between the sill and the lower edge of the sash to raise it to a specific opening angle. For example, a force of about 228 N (51 lbs) would be needed to open a 23 kg (150-lb) emergency exit window to an angle of 40°.

Estimation of the force required for a given opening distance requires knowledge of the emergency window exit dimensions and weight. The dimensional data were gathered during field measurements of window-opening forces, and the weight data were obtained directly from the three motorcoach manufacturers.

Table 3-1 presents motorcoach emergency exit window dimension and weight data in both English and SI units.

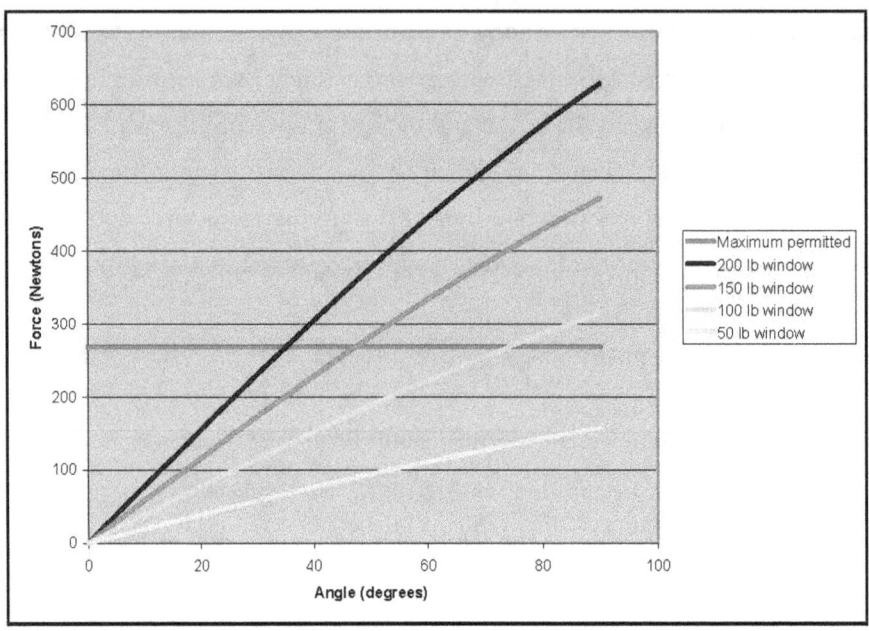

Figure 3-4. Force Required to Open Emergency Exit Window to a Specified Angle

Table 3-1. Emergency Exit Window Dimensions and Weight

MANUFACTURER	MODEL	WIDTH In (cm)	HEIGHT in (cm)	WEIGHT lb (kg)
MCI	D	55 (140)	34 (86)	95 (43)
	J	65 (165)	37 (94)	150 (68)
Prevost	Le Mirage	33 (84)	42 (107)	~80 (36)
	X3-45	60 (152)	37 (94)	94 (43)
	H3-45/41	60 (152)	37 (94)	103 (47)
Van Hool	C2045	65 (165)	34 (86)	155 (70)
	T2145	77 (196)	34 (86)	185 (84)

To achieve a given clear opening for passenger egress, the emergency exit window must be pushed out by an amount equal to the clear opening, plus the amount by which the sash frame and release bar (if present) project into the space between the glass and the sill. The opening angle associated with any given opening distance can be calculated from the height and opening distance using the sine and cosine rules from trigonometry. These calculations are easily accomplished using an applet, such as one found at http://www.saltire.com/applets/triangles/tri3s.htm. For example, to calculate the opening angle for a window 34 inches in height with a 3-in projection from the glass to provide a 20-in clear opening, one would enter the number 34 for two sides of the triangle and 23 for the third side, yielding an opening angle of 39.5° calculated by the applet in the link above.

With the angle thus calculated, the opening force for a 21-inch clear opening and a 150-pound window (68 kg) would be:

$$F' = ((150 \text{ lbs}/2) \times (\sin 39.5°) / \cos (39.5° / 2) \qquad (3)$$

$$= 50.6 \text{ lbs. } (225 \text{ N})$$

C. Actual Release and Opening Forces

At the beginning of the Volpe Center study, it was assumed that emergency exit windows installed on motorcoaches might exhibit some "sticking" problems (documented for passenger rail car emergency windows) that would cause actual opening forces to be substantially greater than the forces using the calculated procedure. Accordingly, actual measurements on various motorcoaches were conducted with a force gauge coupled to an extension, one version of which is shown in Figure 3-5.

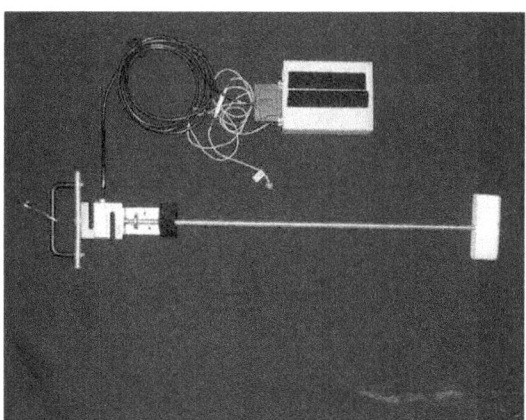

Figure 3-5. Force Gauge Attached to an Extension

This force gauge is the same one used for the release-force measurement. The steel rod running between the S-beam load cell and the wood block can be readily exchanged for others of different lengths. Figure 3-6 shows that the overall length from the "D" handle to the outside of the wood block is 70 cm (27.5 in). In use, this apparatus is simply used as a prop between the exit window sill and the flange at the bottom of the sash, thus holding the exit window open to

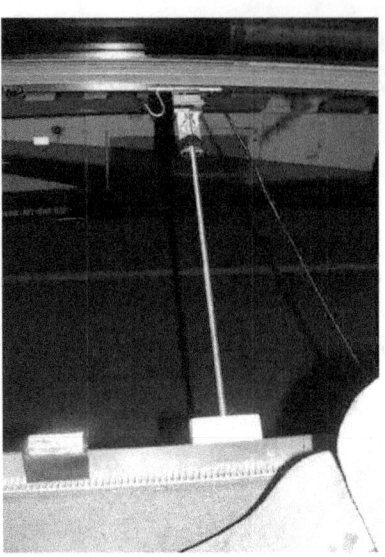

Figure 3-6. Motorcoach Emergency Exit Window Open with Force Gauge

some fixed amount. The 70 cm (27.5 in) version yields a clear opening of 58 cm (23 in) on a 2006 *MCI* "J" model emergency exit window. Note that most of the S-beam load cell and the "D" handle are out of sight beneath the release bar.

Volpe Center staff originally believed that all measurements of the static force needed to hold emergency exit windows open could be made with a gauge of constant length. However, when the measurements began, it became apparent that the different retention mechanisms protruded into the space between the window and the sill by various amounts that made the effective opening size substantially smaller than the gauge length and variable. Furthermore, after the initial emergency exit window egress tests were conducted at the MGA test facility, it became apparent that larger window exit openings might be required than had been previously contemplated. This led to experimentation with gauge assemblies of different designs and dimensions. As a result, the opening forces were measured at various amounts of emergency exit window opening.

The emergency exit window release and opening data results are discussed in Chapter 6, while those for emergency roof exit hatches are reported in Chapter 7.

3.3 MGA Test Facility Field Visit

In conjunction with the NHTSA-contracted motorcoach tests conducted by the MGA test facility, located in Burlington, WI, Volpe Center staff made force measurements for the emergency window exits installed in each bus before the roof crush tests were conducted.

In addition, Volpe Center staff "dropped" out of several open emergency exit windows on each motorcoach, before the roof crush/rollover tests were conducted, as well as the roof hatches of both vehicles, while they were on their side after those tests were completed. (See Section 3.5.)

Volpe Center staff also had the opportunity to inspect the emergency rear door, window exits, and roof hatches of a school bus that had been previously tested according to the FMVSS 220 roof crush tests. Volpe Center staff observations and measurements obtained from the MGA field visit are presented, as applicable, in Chapters 4-8.

3.4 VOLPE CENTER EGRESS EXPERIMENTS

3.4.1 Overview

In the decades since the University of Oklahoma research studies were completed, several significant changes in motorcoach design have occurred, which makes it necessary to update egress rate estimation using current bus design. Motorcoach design changes include:

- An increase in floor height of 30 to 61 cm (12 to 24 in);
- "Kneeling" capability (which lowers the height of the bottom step from 43 cm to 30 cm (17 in to 12 in);
- An increase in window sill height of about 0.6 m (2 ft);
- A large increase in window size and weight;
- Elimination of rear and rear side doors for emergency egress; and
- Introduction of wheelchair-access doors (which can not be opened from inside in current intercity motorcoach design).

Egress rates for anything other than the normal, front-door exit path are highly variable and affected by numerous factors, particularly:

- Fitness of subjects;
- Knowledge of procedures for opening, securing, and traversing exits;
- Lighting conditions;
- Physical orientation of the vehicle; and
- Ground conditions at the end of the egress path.

3.4.2 Experiment Study Design

An *MCI* J-4500 motorcoach was rented from Peter Pan Bus Lines and moved to a location at the rear of the Volpe Center building.

Figure 3-7 shows the motorcoach interior and the exterior, as located on the Volpe Center campus. The bus was equipped with a wheelchair door located on at the rear of the bus next to the lavatory on the right side (see the open door on the rear far left of Figure 3-7b).

a. Interior   b. Exterior

Figure 3-7.   Volpe Egress Experiment Motorcoach

Figure 3-8 shows the bus layout and the camera location for the three experiments.

The type of exits selected for the egress experiment trials conducted included front-door, emergency exit window, and wheelchair-access door, in that order. A large 10-cm-(4 in-) thick polyurethane-foam landing mat was located directly under the emergency window exit and the wheelchair-access door for subjects to drop onto after using the respective exit. Cameras were installed inside and outside of the bus to record the egress actions and times for each trial. Six video cameras mounted at various locations inside and outside the motorcoach provided detailed views of each portion of the egress. The outputs of all cameras were recorded with a common time-stamp (hh:mm:ss.sss format) on a multichannel digital video recorder (DVR) to facilitate precise determination of egress trial timings.

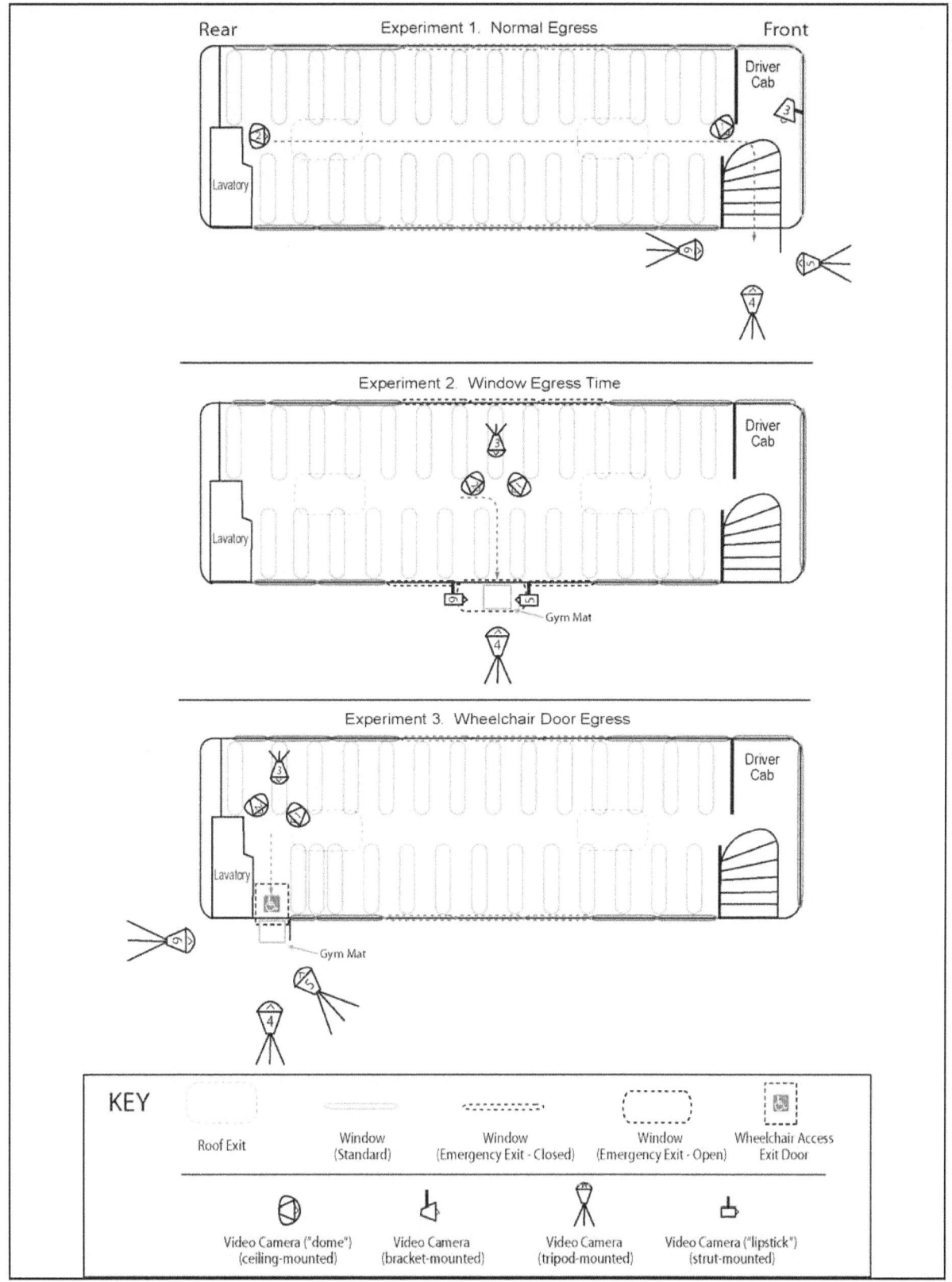

Figure 3-8. Volpe Center Motorcoach Experiment Configuration and Camera Location

3.4.3 Subject Selection

Participation was limited to federal employees of the Volpe Center for the following reasons:

- Security restrictions;
- Health Service nurse was authorized to provide care only for federal employees;
- Subjects received their normal salaries while participating, so no additional compensation was necessary; and
- The Workmen's Compensation Act applied in the unlikely event of a serious injury.

The number of subjects selected for the various trials were as follows:

- 54 subjects for the normal (front-door) egress test;
- 6 subjects for the emergency window exit-egress test (1 short female and 5 other males, including the very tall 6-ft 4-in) subject; and
- Five of the same 6 subjects for the wheelchair-access door-egress test as for the window egress test (except the very tall subject).

Individuals with physical disabilities were excluded from all tests for the following reasons:

- Risk of injury from falls would have been much greater; and
- So much variance in the egress time would be introduced as to mask the effects of design changes.

All of the subjects were briefed by one of the Volpe Center staff investigators using the Institutional Review Board (IRB) protocol prior to the conduct of the experiment trials. All subjects reviewed and signed an informed consent form. For the emergency exit window and wheelchair-access door egress trials, the subjects were physically fit and confident of their ability to perform the respective egress procedure, after watching video illustrations and receiving briefings from Volpe Center staff. Only individuals with sufficient upper-body strength to support their body weights were included as subjects. Potential subjects were free to make a final judgment about whether they had sufficient strength after viewing a video of the emergency exit window egress procedure.

For the 54 subjects participating in the front-door egress trials, an attempt was made to obtain a representative cross section of Volpe Center employees in each of the age and gender categories. The three age groups were: under 35 years, 35-55 years, and over 55. About one half of the total was divided between males and females. Other parameters, such as subject height and weight were not considered during the subject recruitment and selection process.

3.4.4 Experiment Trial Protocols

There were no controls; i.e., all subjects in a given experiment performed the same egress procedure for each trial.

The experiments were conducted on the Volpe Center campus in the afternoon during duty hours. All of the egress trials were completed during a three-hour time period.

3.4.4.1 *Front Door Egress*

No data have been gathered regarding normal, front-door egress rates since NHTSA sponsored research programs conducted in the 1970s, despite significant changes in motorcoach design (higher floors, "kneeling" capability, etc.). An estimate of front-door egress rates is essential for assessing whether successful evacuation can be accomplished in certain particularly hazardous scenarios, such as fires.

The objective of this experiment was to measure flow rate under benign conditions for able-bodied adults exiting from the bus via the front-door. Because the subjects were not encumbered with luggage, parcels, or small children, the egress rates observed were expected to be substantially higher than those observed at bus terminals by typical revenue passengers. The results from the experiment trials were used to provide a basis for estimating the lower bound for total egress time in an ideal "best-case" emergency scenario; i.e., one in which the coach is upright on level ground, there is no fire or smoke, the front door is available, all passengers are able-bodied and uninjured, and the interior and exterior are well illuminated.

The subjects filled every seat on the motorcoach for three of the four egress trials (there was 1 person who did not arrive in time for the first trial). The front door was open for all four trials; before each trial, the subjects first seated themselves on the bus. When all subjects were seated, the Volpe Center staff experimenter gave an oral instruction via the public-address system to exit via the front door "quickly" (simulating that they were slightly late for work) and to move directly away from the bus door after leaving the bus. Because stair pitch is a well-established determinant of walking speed on stairways, the kneeling feature was used for alternate trials. The subjects exited the bus twice with the kneeling feature for the front door steps activated, so that the bottom step was 30 cm (12 in) above the ground, and twice with the front door steps in the non-kneeling position with the bottom step, at a height of 43 cm (17 in) above the ground. These heights were slightly less than measured for motorcoaches on pavement because the subjects in the Volpe Center experiment were stepping onto grass turf about 5-cm (2-in) thick.

For each trial, the subjects seated themselves on the motorcoach and were instructed to choose different seats in each trial, well separated from wherever they sat previously. When all were seated, they received an oral instruction to evacuate via the front door by the experimenter using the public-address system and also signaled with a flag wave. As soon as all subjects were off, they were instructed to reboard for the next trial.

3.4.4.2 Wheelchair-Access Door

The objective of this experiment was to develop a rough estimate of the flow rate through an unobstructed wheelchair-access door of typical current design, when used by able-bodied adults. One of the options being considered to reduce the time necessary to evacuate a bus is to modify the design of motorcoach wheelchair-access doors so that they can be opened from the inside. This would allow egress from a height of 1.5 meters (5 feet), as opposed to 2 meters (7 feet) for the emergency exit windows, and would allow for a substantially higher occupant flow rate than that of those exits.

The primary selection criterion was that the subjects confirmed their ability to drop a distance of three feet without pain or injury. The subjects included four males ranging from 29 to 58 years of age and one 23-year-old female.

The technique used to exit the wheelchair door consisted of a "sitting jump." Some foreign passenger railroads also include pictograms illustrating the technique in their emergency egress instructions, as shown in Figure 3-9.

Figure 3-9. Pictogram Illustrating the Sitting Jump Egress Method

Prior to the start of the experiment, Volpe Center staff demonstrated the "sitting jump" egress technique from the bus floor to the subjects who then directed the subjects to follow as rapidly as possible. Two trials were conducted.

3.4.4.3 Emergency Exit Window Egress

Data are not available that describe the rate at which individuals of normal physical ability can traverse an emergency exit window with a 2-m (7-ft) drop from the sill to the ground. The primary objective of the two experiment trials was to measure the window egress flow rate for able-bodied adults. A second objective was to determine what size opening between the window sash and the side of the coach is required for a +95th height percentile male to pass through it unobstructed.

The primary selection criterion was that the subjects confirmed their ability to drop a distance of 5 feet without pain or injury. The subjects included five males ranging from 29 to 58 years of age and one 23-year-old female. One subject was a 6-ft-4-in male.

Prior to starting the trial, the selected emergency exit window was propped open with a pair of 91-cm- (36-in-) aluminum struts, on which two of the exterior video cameras were mounted. Volpe staff demonstrated the "controlled" drop method to use from the window sill to the mattress pad twice to minimize risk of injury to the 6 subjects.

3.5 EMERGENCY ROOF EXIT HATCH EGRESS EXPERIMENTS

Compared with other means of egress from a motorcoach, the emergency roof exit hatch allows more variety as to method, with each of the various methods placing different demands on the user and imposing different risks. There are at least three methods of exiting through a roof hatch when the motorcoach is on its side:

- Somersault,
- Whole-body lift, or
- Cautious approach.

Before planning any human-subject exit experiments involving the emergency roof exit hatch, Volpe Center research team members investigated each of these means through personal experience at the MGA test facility, after the NHTSA-contracted 2 motorcoach-tip-over tests were completed. The *MCI* bus roof hatch opening was 87 by 52 cm (34 by 21 in), while the *Prevost* motorcoach hatch opening was of 56 by 56 cm (22 by 22 in).

The advantages and risks for each method that can be used for emergency roof exit hatch egress, as used by Volpe Center research team members, are described in the following sections.

3.5.1 Somersault

In this method, the occupant's head and torso emerged from the hatch first. The occupant then bends down as far as possible, extending arms so that they nearly reach the ground and resting body weight on the lower edge of the hatch. With a slight thrust from leg muscles, body center of gravity is shifted outward, while arm, back and hip muscles are used to pull the rest of the body through the opening in a somersault.

While this method is fast and requires relatively little physical strength, it would expose the occupant to a risk of lacerations and bruises, if the surface outside the roof hatch were not smooth and soft, e.g., grass.

3.5.2 Whole Body Lift

This method is similar to that used by ship personnel to traverse bulkhead doors rapidly. It involves the occupant lifting both feet through the opening simultaneously by grasping some part of the bus structure above the hatch with both hands and raising the torso to a height such that the legs clear the lower edge of the roof hatch.

This method requires sufficient upper-body strength to raise one's entire body weight, sufficient finger strength to grasp whatever handhold can be found, the ability to see or feel that handhold, and the presence of such a handhold. Although ship bulkhead doors have handholds, there are no such devices explicitly designed for motorcoach roof exit hatches. During the MGA tests, one person was able to grasp the lip of the roof hatch opening with his finger tips while another found that the opening in the luggage rack where a pair of reading lights had been (they were dislodged by the bus tip-over impact) provided a safe handhold for gloved hands.

3.5.3 Cautious Approach

An occupant unwilling to accept the risks or lacking the strength to use the foregoing methods would probably elect to use a cautious approach, extending only one limb at a time and keeping the body well supported, in order to egress through the emergency roof exit hatch.

3.6 RESULTS

Results for the Volpe Center emergency exit release and opening force measurements, as well as those for each of the Volpe Center-conducted motorcoach egress experiments are discussed in Chapters 4-7.

4. FRONT DOOR EGRESS

The front door exit (also known as a "service door") is the most often means of egress used in most emergency situations involving motorcoaches. The front door allows for a much higher passenger flow rate than the alternatives, as well as much less risk of injury. However, in serious frontal collisions and rollovers, which include many fatal bus crashes, the front door is usually unavailable.

4.1 FMVSS 217 REQUIREMENTS

The scope of FMVSS 217 motorcoach requirements discussed in this study is limited to emergency exits. By definition, front doors are outside its scope, unless they meet those emergency exit requirements. However, since the front door is the primary means of egress from the motorcoach in emergencies, passengers will use it when the bus is upright and the door is operable. The area of the front door opening may be counted toward meeting the required total emergency exit area if it meets all other emergency exit requirements and is marked as an emergency exit. However, since most motorcoaches have greater emergency exit window egress area than FMVSS 217 requires for the total egress area, the marking of front doors as emergency exits by manufacturers is effectively optional for most motorcoach designs.

FMVSS 217 requires that school buses have either a rear exit door that provides an unobstructed opening large enough to fit a rectangular parallelepiped of at least 114 cm (45 in) high by 61 cm (24 in) wide by 30 cm (12 in) deep or an emergency exit side door that provides a minimum opening size of at least 41 cm by 122 cm (16 by 48 in) and have a 305 mm (12 inch) clearance to the aisle.

School bus emergency door releases must be able to be operated by a single person without the use of remote controls or tools, notwithstanding loss of power. In addition, school bus emergency exit doors must be equipped with a positive door-opening device that bears the door weight, keeps the door from closing past a specified point, and provides a release or override, regardless of the bus body orientation. Moreover, no additional action must be required beyond opening the door past the point at which the door is perpendicular to the bus body. An interlock alarm must be audible to the bus operator and in the vicinity of the emergency door exit(s), if the ignition is on and the exits are not closed.

In its "Basic Plan for Motorcoach Passenger Safety Awareness,"[51] FMCSA notes that operators should: "Emphasize that, whenever possible, the motorcoach door should be the primary exit choice" for emergency egress.

4.2 DESIGNS IN USE

As noted in Section 4.1, NHTSA does not consider a motorcoach or other bus front door to be an emergency exit unless it complies with FMVSS 217 requirements.

Figure 4-1 shows the interior and exterior views of a typical newer model motorcoach, with the front door open.

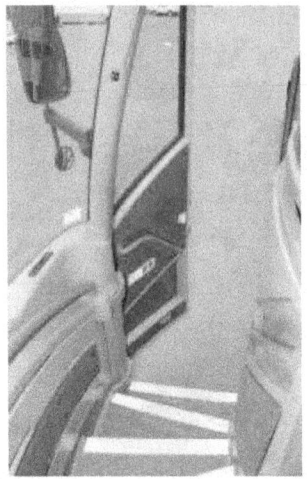  

a. Interior   b. Exterior – No kneeling and Kneeling

Figure 4-1.   Motorcoach Front Door and Stairs

Due to their higher floor heights to accommodate under bus luggage storage, motorcoaches require passengers to use a greater number of steps to get on and off than school buses and transit buses; the latter usually have no more than three steps (see Figure 4-2 and Figure 4-3).

  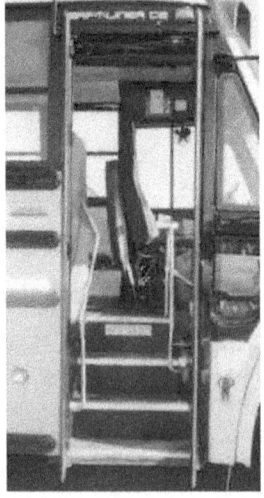

a. Interior – closed and open   b. Exterior – open

Figure 4-2.   School Bus Front Door and Stairs

a. Interior   b. Exterior

Figure 4-3. Transit Bus Front Door and Stairs

Motorcoaches of recent design have power-actuated front doors. A touch of a button or switch on the driver's console is all that is required to open or close them. When the system is functioning normally, pressing the control is the easy way to open them. With the engine off and no air pressure in the actuator, the front door can easily be pushed open, unless the mechanical lock has been set. All motorcoaches have a lock that is key-operated from outside and by some kind of lever or knob from inside (see Figure 4-4). On some motorcoaches, this lock release is labeled and is easily recognized, while on others, it is not labeled and is hard to see, particularly in dim light.

Figure 4-4. Motorcoach Interior Latches for Front Door

The pneumatic systems that operate the doors are designed to retain pressure for many hours after the engine is switched off. Therefore, many modern coaches include an override control to release pressure in the actuator in the event of some malfunction. This manual control is usually red in color (see Figure 4-5) and may or may not be labeled to explain its use.

a. MCI　　　　　　　　　　　　　　　　　　　b. School Bus

Figure 4-5.　Door Pneumatic Actuator Pressure Release Knob and Instructions

Older designs (1980s and earlier) have a mechanical arm linking the front door to a crank within the driver's reach, as shown in Figure 4-6. The front door cannot be pushed open until the crank handle is pulled out of its latched position. Bus riders who have observed drivers performing this action numerous times would presumably know what to do to operate the door; however, no instructions are provided to make the door operation explicit.

Figure 4-6.　Hand-Crank Front Door Opener

Since all of the front doors examined were very easy to open manually (with the actuator depressurized) and effortless under power, no force-gauge measurements were recorded. The principal issue in emergency egress via the front door is not the force required to open it, but rather whether passengers can comprehend quickly how to open the door if the driver is incapacitated. If the front door is available and the actuator is pressurized, but the driver is incapacitated, egress may be significantly delayed because passengers may not know which control operates the door. Some control panels have a switch with an icon that obviously represents the front door, while others (red button in Figure 4-7a) do not have such markings. Furthermore, even if the switch has a label or icon, it may be hard to read due to wear, as shown in the four-year-old bus (upper-right-hand switch in Figure 4-7b).

a. *Van Hool* door control   b. *MCI* door control

Figure 4-7.   Typical Motorcoach Driver Front Door Controls

The time it would take for passengers to determine how to open the front door in an emergency situation is not known; no controlled experiments have been conducted to answer this question. However, some Volpe Center staff members who were asked to open the door needed more than a minute to do so.

Bi-fold doors are often used in school and transit operations to speed loading and unloading, but are not used for motorcoaches. Normal school bus and transit bus front door controls are illustrated in Figure 4-8. (The transit bus front door shown is operated with a hand crank handle, to the left of the steering wheel; some newer buses use a door switch in a similar location.)

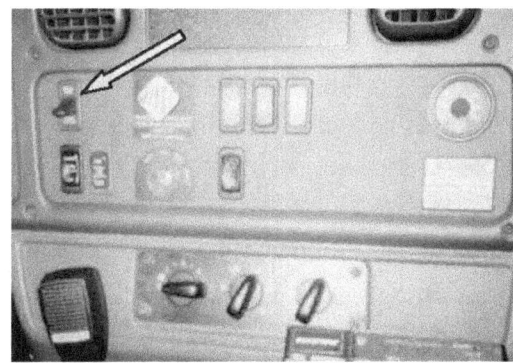 

a. School bus   b. Transit bus

Figure 4-8.   Typical School Bus and Transit Bus Driver Front Door Controls

4.3   USABILITY AND EGRESS RATES

Motorcoach egress rates are affected by numerous physical properties of the type of emergency exit that is available for use by passengers. While extensive literature regarding emergency egress from buildings exists, there is very little research on egress from highway vehicles. As indicated in Chapter 2, the only significant study for U.S. intercity bus emergency egress was completed 30 years ago. Since then, motorcoaches have increased significantly in size and floor height and gained kneeling capability. Accordingly, estimation of current motorcoach egress rates requires data from vehicles representative of the existing U.S. fleet. Volpe Center staff obtained data, for certain key variables, as described in the following paragraphs.

The principal determinant of front-door egress rate is stair pitch. The egress rate for typical passengers walking single file down a level aisle is about 53 persons per minute.[52] Walking speed varies with the pitch of the stairs: the steeper the pitch, the slower the walking speed.[53][54] However, the rise from the pavement to the first step of a motorcoach stairway is much larger than any other step. Therefore, the front-door egress rate is effectively determined by how rapidly passengers can make the last step off the bus.

Many motorcoaches are now equipped with "kneeling" capability, i.e., an active air-suspension system that can be depressurized to lower the normal height of the first step to about 28-30 cm (11-12 in), as opposed to the normal height of 41-43 cm (16-17 in) when the coach is in motion, This capability enables passengers to more easily board and deboard the bus, particularly older persons and persons with mobility impairments. However, even at the reduced height of the kneeling position, the first step height is substantially greater than the normal (up to 20 cm (7.75 in) height permitted for stairways by building codes. Such higher than normal step heights *i*ncrease the risk of tripping and falling. However, the minimum height above the road of the lowest step is determined by the distance required for adequate road clearance, rather than passenger convenience and safety.

The most useful empirical data for estimating motorcoach egress rates must be collected from observing motorcoach egress, because stairs on other vehicles have different dimensions, especially for the lowest step.

Given the limited research data for motorcoach front-door egress, Volpe Center staff decided to measure front-door egress in two ways: naturalistic observations-and controlled experiments. Chapter 3 describes the methodology used for each of the activities.

4.3.1 Naturalistic Observation of Egress Flow Rate Results

The passenger observation data shown in Table 4-1 and Table 4-2 were collected at the Boston South Station bus terminal. Passenger counts and elapsed time data for deboarding for motorcoaches at the Mendenhall Glacier Tourist Center[49] appear in Table 4-3.

The data in Tables 4-1 through 4-3 indicate the following points for motorcoach passenger egress rate using the front door:

Table 4-1. Observed Motorcoach Passenger Egress – South Station, Boston, MA, December 27, 2007

CARRIER	ARRIVING FROM	BAY	TIME OF DAY	NUMBER OF PAASENGERS COUNTED	ELAPSED TIME (sec)	FLOW RATE (ppm)	COMMENTS
Concord Coach	Concord, NH	15	14:21	17	87	12	---
Peter Pan Bus Lines	Springfield, MA	7	14:27	47	215	13	28 sec for first passenger (elderly, impaired) to get off; another disabled (crutch) passenger left a gap of about 25 sec; one passenger with baby in arms. Stopped count and time with a few passengers still on bus with too many parcels to manage.
Fung Wah	NYC	13	14:50	43	90	29	---
Concord Trailways	Portland, ME	14	14:59	4	15	16	---
Lucky Star	NYC	13	15:07	50	179	17	Stopped count with few passengers still on bus with many parcels
TOTALS				161	586	16	---

61

Table 4-2. Observed Motorcoach Passenger Egress – South Station, Boston, MA, May 23, 2008

CARRIER	ARRIVING FROM	BAY	TIME OF DAY	NUMBER OF PASSENGERS COUNTED	ELAPSED TIME (sec)	FLOW RATE (ppm)	COMMENTS
Peter Pan Bus Lines	Hartford, CN	7	13:50	28	95	18	Older demographic, lots of bags
Fung Wah	NYC	25	14:15	36	105	21	One passenger reboarder disrupted flow
Lucky Star	NYC	25	14:20	40	105	23	
Peter Pan Bus Lines	NYC	22	14:30	30	88	20	Lots of bags, bag collectors obstructed flow
Peter Pan Bus Lines	?	17	14:40	38	124	18	=
Omega Express	?	24	15:04	29	76	23	---
Plymouth & Brockton	Hyannis, MA	6	15:22	12	49	15	Each passenger gave something to driver during exit
Concord Coach	Portland, ME	7	15:26	7	32	13	---
TOTALS				220	674	20	---

Table 4-3. Observed Motorcoach Passenger Egress – Mendenhall Glacier Visitors Center, Juneau, AK, August 14-15, 2006 (1)

SPACE / POSITION	TIME OF DAY	NUMBER OF PASSENGERS COUNTED	ELAPSED TIME (hh:mm:ss)	FLOW RATE (ppm)	COMMENTS
1	11:50:30	40	0:02:25	17	Talked to lady getting off and used lift to get wheelchair passenger off
2	14:48:00	36	0:02:23	15	Stuck behind vehicle in space 1, last passenger off bus dawdled ~ 40 sec
3	14:50:50	35	0:01:55	18	Stuck behind vehicle in space 1, last passenger off bus dawdled ~ 40 sec
5	14:53:40	50	0:02:12	23	Parked way back
4	14:53:30	45	0:03:10	14	Parked way back
2	14:58:45	28	0:02:15	12	
3	14:59:06	50	0:02:34	19	
4	14:59:19	45	0:03:06	15	
3	8:30:42	45	0:03:01	15	Driver spoke with us and during that time, a couple of passenger returned to leave bags
3	9:10:20	38	0:01:55	20	No one in front, 1 person went back on bus to retrieve something, relatively young group (agile)
3	9:24:17	52	0:02:54	18	One coach in front
2.5	9:30:00	51	0:03:17	16	5 buses lined up
5	9:35:00	30	0:01:42	18	Buses waiting to turn into drop-off lot
3	9:52:45	20	0:01:00	20	
2	10:00:39	50	0:03:37	14	
2.5	10:09:28	32	0:01:54	17	Needed to do 3-point turn to getting into lot, backed out to get around vehicle in position 1.5
2	8:28:06	49	0:02:34	19	Long delay - the driver was chatting with passenger

Table 4-3. Observed Motorcoach Passenger Egress – Mendenhall Glacier Visitors Center, Juneau, AK, August 14-15, 2006 (2)

SPACE / POSITION	TIME OF DAY	NUMBER OF PASSENGERS COUNTED	ELAPSED TIME (hh:mm:ss)	FLOW RATE (ppm)	COMMENTS
	8:40:48	51	0:02:56	17	Pedestrian congestion at trailhead
	9:22:05	28	0:04:51	6	Passenger went back for water bottle; driver took photos of bears
4	9:26:00	42	0:02:17	18	Bear sign is a popular photo op
5	9:38:28	39	0:01:28	27	It was clear at front - no need to be in space 5
	9:58:30	52	0:03:57	13	One kid needed a wristband
	10:03:56	28	0:01:07	25	Had to get strollers out
	10:08:30	35	0:01:34	22	Waited to pull up before offloading; had to draw a map for visitors
	13:15:30	9	0:04:35	2	
	13:20:10	38	0:02:20	16	
2	13:07:04	47	0:01:56	24	
	13:44:41	35	0:02:46	13	One passenger went back to retrieve an item
1	14:08:43	34	0:01:42	20	Wheelchair
	14:24:14	33	0:01:44	19	
3	14:32:10	55	0:03:05	18	Passed out wristband before offloading
1	14:47:46	46	0:02:11	21	Passenger went back for coats
	14:57:30	30	0:01:51	16	Parked right by trailhead
2	15:13:16	36	0:02:27	15	Wheelchair
TOTALS		1334	1:24:41	16	

- The only previously known controlled measurements of front-door egress rate from a motorcoach in the United States were made by the OKRI studies.[19][24] The 1970 OKRI data indicated that 8 subjects exited by the front door within 21.8 seconds, which can be extrapolated to a flow rate of 22 persons per minute (ppm). The 1978 OKRI data indicated that 19 subjects exited via the front door in simulated darkness (using dark goggles). The subjects took 9.8 seconds to open the door and 77.6 seconds to complete the egress, from which a flow rate of 17 can be calculated. Mean egress rates ranged from 16 to 20 ppm. The fastest observed rate was 29 ppm.

  The lowest observed egress rates, which occurred at the Mendenhall Center, were likely not related to any physical difficulty in egress, but rather to the fact that passengers were tourists. They had no reason to rush, and had to put their coats on and collect their cameras, binoculars, etc., in the process of deboarding.

- Within a given busload of passengers, clusters of agile, unencumbered passengers followed each other so closely that the egress rate exceeded 40 persons per minute.

- A fully loaded motorcoach can be unloaded in less than two minutes if all passengers are able-bodied and unencumbered by unwieldy carry-on packages or luggage.

- Passengers with mobility impairments are often seated near the front of the bus so that the driver can most easily and quickly assist them in getting on and off. Individuals needing such assistance created gaps in the flow of about 30 seconds each. The impaired persons observed required a crutch or a cane. No wheel-chair users were observed.

- The presence of mobility-impaired passengers and one or two parents with small children can slow the mean egress rate substantially. Total egress time can be roughly estimated to increase by about 30 seconds for each impaired passenger and 15 seconds for each passenger with small children or carrying unwieldy parcels.

The *TCRP 100 Manual* containing transit capacity estimates provides an additional source of naturalistic observations.[48] The TCRP summary of door flow observations for various U.S. transit operations involving high-floor buses indicates that the average time per alighting passenger is about 3.3 seconds for a single-channel doorway. This is equivalent to 18 ppm, essentially identical to the average of the Volpe Center staff observation at the Boston, MA bus terminal, as shown in Table 4-1 and Table 4-2.

4.3.2 Volpe Center Experiment Front Door Egress Flow Rate Results

As described in Chapter 3, Volpe Center staff conducted a front door motorcoach egress experiment (see Figure 4-9). The experiment objective was to collect "baseline" exit times under controlled best-case conditions. The experiment included four trials: two in which the bus was in the kneeling condition, and two in which it was not. The subject group was comprised of Volpe Center federal employees, including a balanced mix of genders and age groups.

Figure 4-9. Volpe Center Motorcoach Experiment to Determine Front Door Egress Rate – Best-Case

The data from the four front door experiment egress trials are presented in Table 4-4. The mean flow rate for the four trials was 36 ppm, and the differences in the mean flow rate among trials are not significant. There were no injuries to any subject participants, or any unusual behavior.

Table 4-4. Volpe Center Motorcoach Front-Door Egress Experiment Results

TRIAL	MOTORCOACH CONDITION	SUBJECT COUNT#	START TIME	FINISH TIME	ELAPSED TIME	FLOW RATE (ppm)
1	Kneeling	53 *	14:17:35	14:19:00	0:01:25	37
2	Non-kneeling	54	14:25:25	14:26:58	0:01:33	35
3	Kneeling	54	14:30:42	14:32:12	0:01:30	36
4	Non-kneeling	54	14:39:40	14:41:12	0:01:32	35

One subject arrived late

4.4 REQUIREMENTS FROM OTHER U.S. MODES

Because sufficient capacity in front (or service) doors is essential to the business of transporting passengers, little regulation is required for their size and number. Most regulations for service doors in other major transportation modes are not applicable to motorcoaches, because passenger aircraft, passenger rail cars, and ships are generally much larger in capacity than motorcoaches. However, regulations from other major modes do have explicit door "ease-of-use" requirements.

Principal requirements are:

- *Aviation* – 14 CFR, Subsection 25.811[29] – FAA requires an opening device physically located on each exit with extensive specifications for signage and instructions. Also requires that signs and instructions be illuminated or self-luminous.

- *Rail* – 49 CFR, Subsection 238.235[30] – FRA requires that power-actuated side doors for passenger rail cars be equipped with a manual override device that permits opening the door from inside without power, and that is adjacent to the door it controls, and that allows use of the device without an implement. (Door size must comply with ADA regulations.)

  49 CFR, 239.107[30] requires that doors for emergency egress be lighted or conspicuously marked and that clear and understandable signage be posted explaining how to open the door.

- *Marine* – 46 CFR, Subsection 116.500[35] – Requires that any means of escape must be capable of being opened by one person from either side in light or darkness.

Chapters 7 and 8 more extensively describe emergency door sign marking and lighting requirements.

4.5 INTERNATIONAL BUS REQUIREMENTS

ECE 36[38] (which applies only to single-deck buses) and ECE 107[39] (which applies to both single and double-deck buses) require that each bus be equipped with two service exit doors or one service exit door plus one emergency door exit. The service exit doors are required to be a minimum height of 165 cm (65 in) in intercity coaches and a minimum width of 65 cm (26 in). Emergency exit doors must be at least 125 cm (49 in) high and 55 cm (22 in) wide. Service doors are explicitly allowed to be considered as emergency exit doors. The ECE regulations also require that service doors be easy to open from inside and outside when the vehicle is stationary, and contain extensive design requirements to protect passengers from unintentional opening and from being injured by the power actuators.

ADR 44/02[41] requirements are almost identical to the ECE regulations. However, ADR 44/02 does not contain the same extensive ease-of-use and protection-against-opening-in-motion language of the ECE regulations. ADR 58/00[42] allows smaller door dimensions (137.5 by 55 cm (54 x 22 in)), and makes no mention of ease of use or protection against opening in motion.

British Regulation No. 257[43] contains extensive requirements regarding the dimensions of doors and steps as well as design specifications, e.g., non-slip treads. Requirements vary with seating capacity, single vs. double deck, etc. Minimum height requirements range from 137 cm (54 in) to 152 cm (60 in). Width minima range from 46 cm (18 in) to 53 cm (21 in). Additional provisions require that power-actuated doors be readily operable in the event of loss of engine power, that doors be prevented from opening while the vehicle is in motion, and that passengers be protected from injury by the doors.

4.6 RELEVANT RESEARCH

In 2006, Volpe Center staff conducted several commuter rail car egress flow rate tests using the side steps to a low-level platform.[55] The average flow rate for unencumbered, able-bodied subjects was 41 passengers per minute. Five trials were conducted using 17 Volpe Center employees and the commuter rail cars illustrated in Figure 4-10. The distance from the bottom step to the ground was 39 cm (15.5 in), and the rise of the other four steps was 22 cm (8.5 in).

Figure 4-10. Volpe Center Commuter Rail Experiment
to Measure Normal Egress Rate

As previously noted, in an egress experiment conducted by the Oklahoma Research Institute (OKRI) in 1972.[23] Eight subjects exited via the front door within 21.8 seconds, equivalent to a flow rate of about 22 passengers per minute. In the 1978 OKRI experiments,[24] 19 subjects passed through the front door in 68 seconds during the Test 2 (simulated darkness), from which a flow rate of 17 passengers per minute can be calculated.

IKARUS, a Hungarian bus manufacturer, conducted two front-door egress trials on its Model 256 intercity motorcoach in 1984 using 45 adult subjects, all aged between 20 and 45 years. They completed the egress trials in 37 and 40 seconds.[56] The imputed egress rates are 68 and 73 passengers per minute. The informal report that describes those tests does not provide detail on subject selection criteria or motivation. The relatively young adult subjects may have been employees of IKARUS, and may have been told that IKARUS wanted to demonstrate rapid egress (as is the case in aircraft evacuation-certification trials). The floor height of the IKARUS 256 bus above the ground is specified as 94 cm (37 in), which compares with a floor height of about 140 cm (55 in) on a typical U.S. motorcoach. At such a low floor height, some agile young subjects will simply jump the stairs, which speeds egress substantially.

4.7 DISCUSSION

When available, the motorcoach front door provides the fastest and safest means of egress. Table 4-5 summarizes known data for side door egress under various conditions compiled from several research studies.

The principal human-factors concerns related to front-door emergency egress have to do with scenarios in which the driver is incapacitated. In such situations, egress may be significantly delayed by the time it takes for passengers to identify which control operates the front door (if the actuator is functional) or how to depressurize the actuator (if it is not functional). Observations of Volpe Center research team members who tried to open the different motorcoach front doors suggest that this process can take longer than the time required for an entire busload of passengers to pass through the door. Such signage as currently exists to explain how to open the front door is often placed where it may not be noticed. Furthermore, some observers may find it difficult to interpret.

If motorcoach front doors are to be used for emergency egress, a manual release on or near the door interior, such as that shown in Figure 4-5 for the front door on a school bus, would assist passengers in opening the door.

Since FMVSS 217 applies only to emergency exit marking and instructions, it does not include requirements for front door instructional signage to explain how to perform release and open the front door, unless the manufacturer specifically marks this door as an emergency exit. However, enabling passengers to open the front door quickly to exit in an emergency without exposing them to the risk of injury from unintended openings is a complex issue. The solution relies on both better signage (further discussed in Chapter 8), as well as in more thoughtful placement of emergency exit release controls and their design for maximum crash-worthiness.

Providing passengers with control of the motorcoach front door for emergency egress would necessitate additional measures (interlocks and alarm) to prevent unintended opening in motion, as well as passenger crush injuries (similar to requirements on elevators, subways, etc.).

Table 4-5. Summary Observations of Front-Door Egress Rates

SOURCE	YEAR	SUBJECTS	FLOW RATE (ppm)	
			Experiment	Natural
OKRI	1972	Young-middle-aged adults (n=8)	22	–
OKRI	1978	Typical riders; chose front door; darkness (n=19)	17	–
IKARUS (bus manufacturer)	1984	Hungarian adults, 20-45 years old (n=45)	68-73	–
TRB/TCRP: TC & QS Manual	2003	Revenue passengers alighting from high-floor buses	–	18
Volpe Center	2006	20-65 year old Volpe Center employees, commuter rail car with steps (n=17)	40	–
Volpe Center	2006-2008	Revenue passengers at bus terminals in AK and Boston (n=1715)	–	16-20
Volpe Center	2008	20-79 yr old Volpe Center employees in controlled bus experiment, MCI J4500 (n=54)	36	–

Because ECE regulations for European motorcoaches are design requirements, they impose an extensive list of provisions aimed at prevention of injuries to passengers by power-actuated doors, and specifically require that power-actuated service doors be fitted with manual opening mechanisms that function with or without power and that can be clearly seen and identified by a person standing in front of the door. As noted previously, FMVSS 217 requires school bus emergency exit door alarms and interlocks. However, no industry standards or other requirements apply to U.S. motorcoaches.

4.8 CONSIDERATIONS

NHTSA already requires that front doors intended for use as emergency exits or included by manufacturers to comply with the minimum required surface opening area be marked as emergency exits. In addition, motorcoach operators must provide passengers with information relating to front door use as the primary emergency exit in order to comply with FMCSA training guidance.

Potential design motorcoach changes that may increase the passenger egress rate via the front service door and reduce the risk of injuries during emergencies include:

- Making door-opening instructions / pictograms more conspicuous to increase visibility to passengers if the driver is incapacitated and there is a loss of power.

- Placing the interior door-opening emergency release on or very near the door. The signage explaining how to open the door should always be placed on the door.

- Providing a purely manual means of opening the door by installing:
    - An interlock to prevent unintended opening while the motorcoach is in motion. Prevention of crush or pinch injuries would require sensors and interlocks similar to those required on school buses, and used on transit buses and passenger rail cars.
    - Alarms to notify the driver of an unlatched door or open during operation.

5. WHEELCHAIR-ACCESS DOOR AND OTHER SIDE DOOR EGRESS

Under provisions of the Americans with Disabilities Act (ADA), the Secretary of Transportation issued a final rule in 1998 requiring the installation of wheelchair-access doors and lifts in over-the-road buses, i.e., motorcoaches used for common carriage. These devices (see Section 5.2) are intended to make motorcoach transportation accessible to persons who use wheelchairs. The ADA rule became effective for large intercity carriers on October 30, 2000, and for small operators on October 29, 2001. Since then, many new motorcoaches sold in the U.S. have been equipped with wheelchair-access doors and lifts. Each of the major motorcoach manufacturers offers these lifts in both new vehicles and as retrofits. There are also independent firms that install retrofits. The proportion of the fleet equipped with wheelchair-access doors is increasing rapidly. Federal law requires that 100 percent of the fleet of large intercity fixed-route carriers be equipped with wheelchair lifts by October 29, 2012.

5.1   FMVSS 217 REQUIREMENTS

FMVSS 217 requires that buses other than school buses, provide emergency exits that meet the requirements of Subsection 5.2.2 or 5.2.3, the latter being the school-bus requirements. Subsection 5.2.2 requires side emergency exit windows and at least one rear emergency exit door, but permits a rear engine bus to have, instead of a rear exit door, at least one emergency roof exit hatch. The rear-engine-with-roof-hatch option has been employed in virtually all motorcoaches manufactured in the last 20 years for the U.S. market.

The door area equipped with a wheelchair-access lift on school buses may be counted toward the additional required emergency door exits, if the lift folds or stows in such a manner that the area is available for persons not needing the lift. In addition, FMVSS 217 includes the following school bus wheelchair anchorage requirements:

- A 30-cm (11.8 in) minimum clear opening between the seats adjacent to the door must be provided (see Figure 5-1);

- No wheelchair anchorages are permitted in the area immediately adjacent to the side emergency exit door) (see Figure 5-2); and

- Prohibition of wheelchair anchorages in an area extending 305 mm (12 in) forward from the rear edge of a school bus side emergency door.

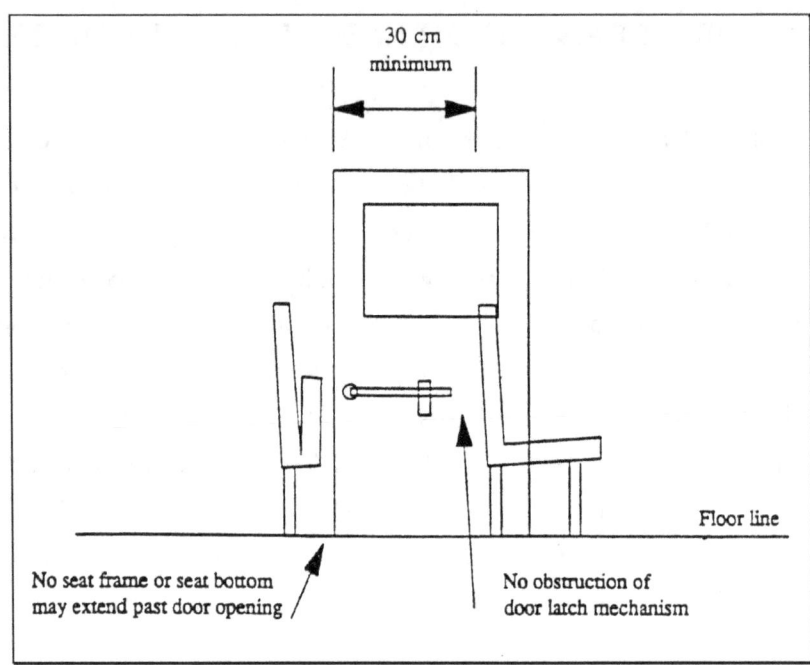

Figure 5-1. FMVSS 217 Clearance Requirements for School Bus Side Door Emergency Exits

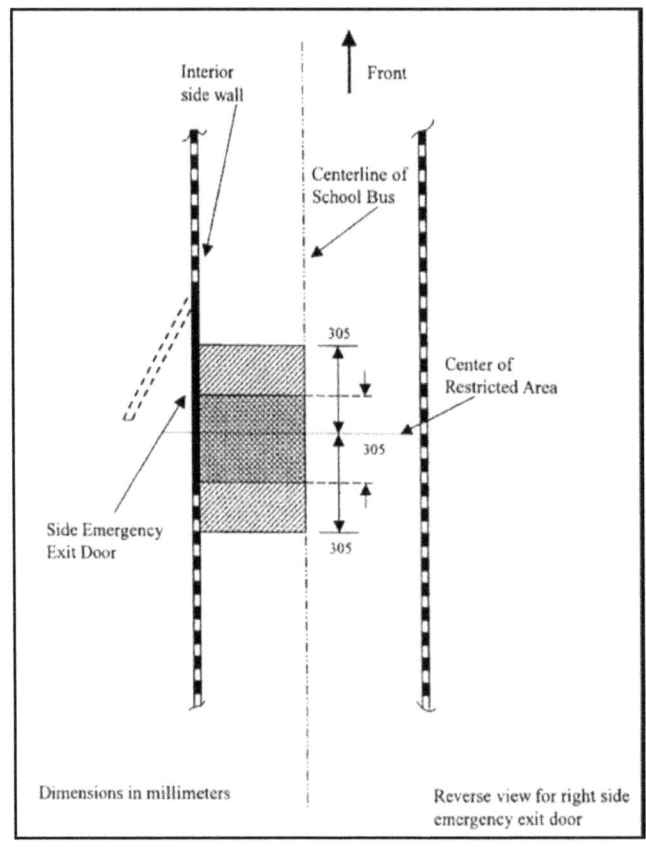

Figure 5-2. FMVSS 217 School Bus Wheelchair Anchorage Location Prohibition

5.2 DESIGNS IN USE

As currently designed, wheelchair-access doors are intended solely for loading and unloading wheelchairs, and cannot be opened from the inside of the motorcoach, school bus, or transit bus. The door release mechanism is located in the compartment below the door where the lift is stowed. This compartment is locked and is normally opened only by the driver. This arrangement prevents accidental opening by passengers inside the bus.

Many intercity and transit bus operators have at least some of their motorcoaches equipped with wheelchair-access doors and lifts and many school buses and transit buses are so equipped.

Figure 5-3 shows the interior location of two motorcoach wheelchair-access doors with the doors in the open and closed position.

a. Rear Side (door open)   b. Mid Bus Side (door closed)

Figure 5-3.   Motorcoach Wheelchair-Access Door Location Interior View – Examples

Figure 5-4 shows the exterior of several motorcoach wheelchair-access doors with the lifts in the open position.

Figure 5-5 shows the interior and exterior of a school bus wheelchair-access door with the door and the lift in the closed position that complies with FMVSS 217. Figure 5-6 shows the interior and exterior of another school bus wheelchair-access door. However, this type of wheelchair-access door does not comply with FMVSS 217 emergency exit door requirements.

Figure 5-7 shows an example of a transit bus wheelchair-access side door and ramp located at the rear side door of the bus. The three steps are used by able-bodied passengers; newer "low floor" buses have the wheelchair access function integrated into the bus front door steps. (There is an emergency release provided for the rear side door (not shown); however, the release is accessed by the use of a small hammer to break the protective glass cover.

a. Ricon (Wabtec) Bay   b. Ricon (Wabtec) Mirage   c. Stewart & Stevenson

Figure 5-4.   Motorcoach Wheelchair-Access Door / Lift Exterior View – Examples

 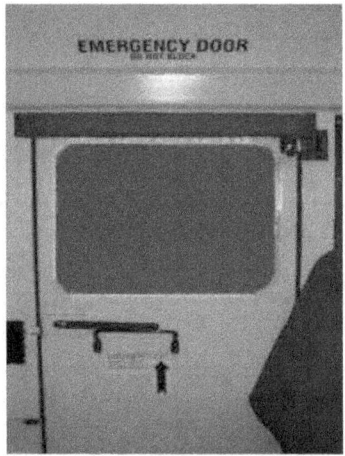

a. Exterior   b. Interior

Figure 5-5.   School Bus Wheelchair-Access Door Emergency Exit – Closed Position

a. Exterior   b. Interior

Figure 5-6.   School Bus Wheelchair-Access Door – Closed Position

a. Interior – Door Closed   b. Exterior – not deployed and deployed

Figure 5-7.   Transit Bus Rear Side Wheelchair-Access Door Location – Example

However, motorcoaches equipped with a rear side wheelchair-access door are designed so that the seat bottom or entire seat located adjacent to that door can be either moved away from the door or removed to provide clearance. For a motorcoach with a wheelchair-access door exit located between seat rows, the clearance varies over a wide range, depending on the degree to which the seats in the forward row are reclined and whether the seat bottom of the rear row is up or down. Clearance should be measured as the distance between vertical planes that are tangent to the rearmost point on the forward row of seats and the foremost point on the rear row of seats, as illustrated in Figure 5-8, derived from a 2004 FAA research report,[57] which describes aircraft over-wing emergency exit access clearances.

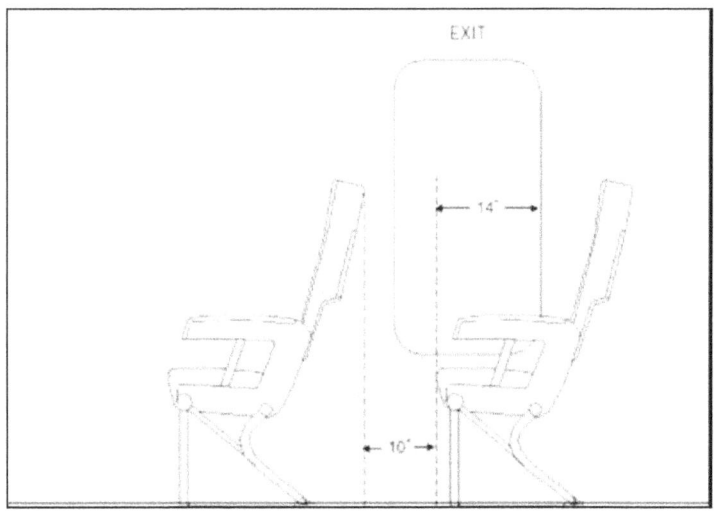

Figure 5-8.   Seat Configuration with a 25.4-cm (10-in) Clearance (Aircraft) – Example

Table 5-1 shows the range of clearances measured in the row with the wheelchair-access door for one *Prevost* X345 motorcoach.

Table 5-1. Wheelchair-Access Door Row Clearances for a *Prevost X-345* Bus

FRONT ROW POSITION	REAR ROW POSITION / CLEARANCE cm (in)	
	Seat Bottom Down	Seat Bottom Up
Upright	29.2 (11.5)	57.8 (22.75)
Reclined	11.4 (4.5)	40 (15.75)

5.3 USABILITY AND EGRESS RATES

To inform decision-making about whether motorcoach wheelchair-access doors or other types of side emergency exit doors should be redesigned to allow their use for emergency egress, it is necessary to estimate how rapidly passengers could exit through the door under various conditions. The simplest and safest condition for egress through the wheelchair-access door is one in which the door is completely unobstructed, all persons are able bodied, and there is adequate illumination. This condition was tested and explained in this section. The Volpe Center will undertake testing of more challenging conditions, such as partial obstruction of the door with seats, in Year 2 of this study.

The Volpe Center conducted a pilot experiment to estimate the time required to open a typical wheelchair-access door of current designs and the egress flow rate through that unobstructed type of door.

5.3.1 Movement of Seats Away from Wheelchair-Access Door

The seats located adjacent to the motorcoach wheelchair-access door were designed so that the entire seat assemblies could be moved forward on floor tracks away from the door to provide clearance. An experienced driver for Peter Pan Bus Lines required 52 seconds to move the seats away from the wheelchair-access door, 12 seconds to walk through a clear aisle to get outside the bus and back to the wheelchair-access door exterior, and 19 seconds to open the wheelchair door and lock it in the open position. The driver needed at least 83 seconds to provide an unobstructed egress path through the wheelchair-access door. If passengers were blocking the aisle, the driver's time to get out through the front door would have been much longer.

5.3.2 Wheelchair-Access Door Experiment Egress Flow Rate Results

The egress rate of 5 subjects using the wheelchair-access door to exit the motorcoach was measured, after moving the seats and locking the wheelchair-access door in the open position.

The sitting jump method of egress used by the subjects is illustrated in Figure 5-9. In the first photo, the subject sits on the door sill and uses both hands to slide forward. Hands are kept on the door sill as long as possible into the fall to retard acceleration and minimize free-fall drop, as shown in the second photo. In both egress trials, egress time for all five subjects totaled 12 seconds, equivalent to a flow rate of 25 ppm.

Figure 5-9. Volpe Center Subjects – Sitting Jump Using Wheelchair-Access Door

This flow rate estimate is not representative of typical motorcoach passengers since all of the experimental subjects were self-selected to have sufficient bone strength and agility to perform the required action without injury concern. Heavy, frail or disabled passengers, who represent a substantial portion of the traveling public, would require the assistance of other passengers or emergency responders.

5.4 REQUIREMENTS FROM OTHER U.S MODES

Wheelchair access is provided through normal service doors in other transportation modes. In some cases, lifts are being added to commuter rail cars that serve stations with low platforms, but no special wheelchair-access doors are being installed in the other modes.

Current FAA requirements for emergency exit row clearances depend on whether a row has two or three seats. For the more common three-seat configuration, the required clearance is 51 cm

(20 inches). However, for smaller aircraft with only two seats between the aisle and the exit, the requirement is only 25 cm (10 inches).[29]

5.5 INTERNATIONAL BUS REQUIREMENTS

ECE 36[38] does not specifically address wheelchair-access doors, but requires that buses have two doors, one of which must be a service door. One of the doors is normally placed in the rear half of the coach (with 4 steps and 5 risers) on the same side as the front service door, and is generally used as a service door to speed loading and unloading. (See Figure 5-10.)

Figure 5-10. MAN Motorcoach –- Middle Side Service Door

The minimum aperture dimensions for emergency doors are 125-cm high by 55-cm wide (49 by 22 in). The free space between the gangway and the emergency door aperture must permit free passage of a vertical cylinder 30 cm (12 in) in diameter and 70-cm- (27-in-) high from the floor and supporting a second vertical cylinder 55 cm (22 in) in diameter; the aggregate height of the assembly of 140 cm (55 in). (See Figure 5-11.)

Additional significant ECE 36 technical requirements for bus service doors used as emergency exits are summarized as follows:

- Shall be easily opened from inside and outside while vehicle is stationary;
- Shall not be power-operated or of sliding type;
- Hinges forward with 100° minimum opening angle; and
- Fitted with audible warning device to warn driver if not securely closed.

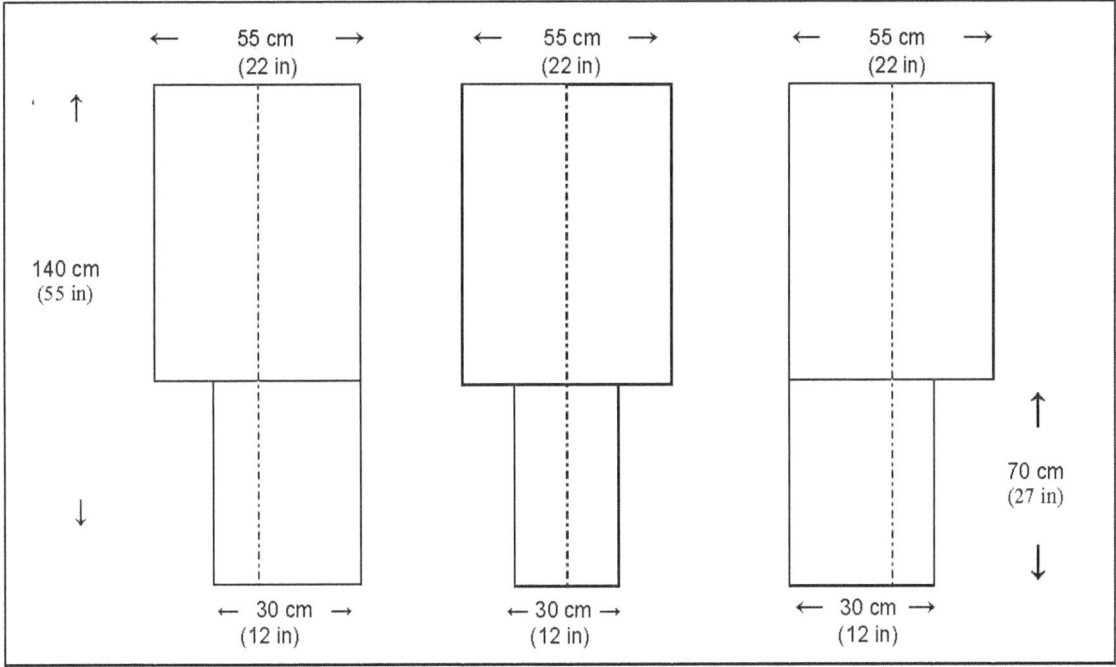

Figure 5-11. ECE 36 – Access to Emergency Doors (Annex 3, Figure 2)

ECE 107[39] requires that a requirement that the door opening control must be placed between 100 and 150 cm (39 to 59 in) above the bottom of the door and within 50 cm (20 in) from the door.

Annex 8 to ECE 107 contains extensive requirements for wheelchair-access doors. Lifts are allowed, but not required since the lower level on some European buses is low enough to be accessible via a ramp and the side service door, provided that door has a minimum width of 90 cm (35 in).

British Regulation No 257[43] does not mention wheelchair-access doors, but does require an emergency door with minimum dimensions of 121 cm (48 in) in height and 53 cm (21 in) in width.

5.6 RELEVANT RESEARCH

Side-rear emergency exit doors were once common on U.S. intercity motor coaches, and egress rates for those types of buses were determined during the OKRI experiments conducted in 1972[23] as illustrated in Figure 5-12. OKRI reported egress rates from the side-rear door as high as 8 persons in 21.8 seconds — equivalent to a flow rate of 22 ppm. This rate was achieved by subjects performing a standing jump from a height of only about 1 m (3 ft) above ground.

Figure 5-12. Subjects Jumping from *GM* PD4104 Bus Rear Emergency Exit (OKRI 1972)

FAA has conducted extensive human-subjects experiments that involved about 2,500 subjects to determine egress through passenger over wing emergency exit openings that are partially obstructed by adjacent seats.[58][59] The spacing requirements for exit-row seats affect the number of rows of seats that can be installed in a particular aircraft, and thus affect the profitability of its airline operation. FAA experiments used such large numbers of subjects in order to achieve sufficient statistical power in the analyses of the factors affecting egress rates.[60] The FAA found that there was no significant improvement in egress rates for increases in the clear opening size of the exit row seats beyond 33 cm (13 in). However, the method of egress for an over-wing exit (stepping through an opening onto the wing) is quite different from that used by passengers to exit from a floor-level door on a motorcoach. Specifically, the sitting jump method of motorcoach egress may become more difficult or require more time to perform as the opening narrows. While a standing jump would still be possible, the free-fall distance to the ground would be greater. Passenger hesitation due to fear of jumping from the height of a motorcoach floor could slow egress.

5.7 DISCUSSSION

In both the OKRI experiments of the 1970s and the recent pilot experiments at the Volpe Center, able-bodied subjects were able to egress through an unobstructed rear side door without steps, at rates that would allow the evacuation of a fully-loaded motorcoach in less than 3 minutes. These findings suggest that the wheelchair-access door could provide passengers with an additional or alternate means of emergency egress that is almost as fast as the front door if it is redesigned for such use in an emergency. It may be feasible for manufacturers to provide an internal emergency release for the wheelchair-access door for use by motorcoach passengers as an alternative to floor-level emergency exits, such as those found on most motorcoaches operated in other countries, if appropriate safeguards and interlocks are also incorporated into the design.

The potential use of wheelchair-access doors for motorcoach emergency egress raises the question about how much, if any, additional clearance space should be required between seats adjacent to the door. While the current FMVSS 217 school bus requirement is 30 cm (12 in) for the space between the seat adjacent to a side-door emergency exit and the seat immediately in front of it, motorcoach adult passengers are often much larger than school children.

The FAA research finding that a 33-cm (13-in) clearance could provide an equivalent level of safety to a 51-cm (20-inch)-clearance has not yet been incorporated into FAA regulations.

Because a wheelchair-access door swings outward from the bus interior, a 13-inch clearance could have applicability to motorcoaches, at least in the case of exits located between seat rows.

It is important to recognize that a given clearance measured by the method illustrated in Figure 5-8 for aircraft seats translates into a much larger opening at torso height for an exit located between seat rows. The same clearance measured between a seat back and a solid lavatory wall would prove restrictive to passengers with a large waist size.

Before determining that motorcoach wheelchair-access doors could be reconfigured by manufacturers for ambulatory passenger emergency egress, the following questions must be answered:

- How rapidly can non-injured adult passengers egress from a wheelchair-access door?
- How does the door placement (between seats or between last row of seats and toilet) affect passenger egress rates?
- How does the egress rate through a wheelchair door compare with a true emergency door, such as, those found on school buses or foreign motorcoaches?
- Are hand and/or footholds needed?
- What existing FMVSS 217 school bus requirements for wheelchair-access doors could be applied to motorcoaches?

These questions are being addressed in the second year of the Volpe Center study.

The ECE emergency door access clearances of 30 cm (12 inches) at the floor and a larger 55-cm-(22-in-) wide opening, 70 cm (28 in) above the floor addresses the fact that motorcoach passengers are adults who may need additional space to be able to maneuver to the emergency exit door and exit the bus.

Motorcoaches with an additional rear-side service door are commonly used for single level passenger seating in Europe. In addition, "tour buses" with two seating levels and substantially higher seating capacities are also widely used. Bi-level bus operation advantages in terms of lower costs and energy consumption per seat mile has created a demand for the introduction of such vehicles into the U.S. market. Both single-level and bi-level coaches have a side-service exit door near the rear because that door is required by both EC 36 and ECE 107, and because it is economically justified for speeding the boarding / deboarding of passengers. That second door is considered to be an emergency exit, and easily meets the required ECE requirements for such doors. However, the motorcoach seating configurations that are currently in wide use require that the rear stairs have a pitch that is substantially steeper than the front steps. No reports of controlled experiments to measure egress rates from such bus doors have yet been identified.

Manufacturers could provide these alternative configurations for U.S. motorcoach operations that use either single or bi-level buses, with the inclusion of a second service side door in the middle or rear half of the bus and a stairway connecting it to the interior seating level. This door could also be used by passengers as an emergency door exit, if it is equipped with an emergency release that is marked and provided with instructions for its use.

5.8 CONSIDERATIONS

As currently installed, motorcoach wheelchair-access doors are of little use for rapid emergency egress because they cannot be opened from inside by passengers (or the driver). Yet, even in their current implementation, after wheelchair-access doors have been opened by the driver or an emergency responder from the outside, they can provide a supplemental emergency access / passenger egress path.

Due to their potential value as another emergency egress route, future wheelchair-access doors installed on motorcoaches could be reconfigured to be operable by passengers from the inside and usable for emergency egress. Such a design change would include:

- Operation of the emergency exit door release from the inside and outside;
- Operation without use of special tools or remote controls;
- Interlocks and alarms to prevent the door from being opened with the bus in motion and to warn the driver whenever the wheelchair-access door is not latched in the closed position; and
- Marking and instructions for operation from both the inside and outside.

As an alternative to the use of wheelchair-access doors for emergency egress, additional side service doors could be installed that could be used as emergency exits. The door releases should also operate and be marked for use in an emergency as indicated above for wheelchair-access doors.

Volpe Center staff are conducting additional human factors egress experiments during the second year of this study. These experiments will use a bus mockup to measure the egress flow rates for a higher number of subjects, with greater gender and age variability for the following types of side doors:

- Wheelchair-access door
    - Measure subject egress flow rates for exiting from the door.
    - Determine how much more additional space, if any, should be required beyond the 30 cm (11.8 in) currently required by FMVSS 217 for school bus wheel chair anchorage locations adjacent to wheelchair-access doors.

- Additional side service door stairway
    - Measure egress rates by subjects using another stairway, other the front door stairway, similar to some international buses, using the same subjects as above.

6. EMERGENCY EXIT WINDOW EGRESS

If the front door is unavailable in a crash or other emergency situation involving an upright motorcoach, the emergency exit windows provide the next best means of escape for the current U.S. fleet. However, survivors of such situations have reported considerable difficulty in opening the emergency exit windows. This issue is highlighted in the 2007 NHTSA research plan.[3] NTSB also recommended that NHTSA consider requiring design improvements to make motorcoach emergency exit windows easier to release and open, as well as provide a means to have the window exits stay in the open position during passenger emergency egress.[8]

6.1 FMVSS 217 REQUIREMENTS

The significant parts of FMVSS 217 with respect to motorcoach emergency exit window egress are summarized below: Total unobstructed area of exit in square centimeters must be at least 432 times number of seats with:

- At least 40 percent of required area on each side of bus;
- No one exit can count for more than 3,458 cm^2 (536 in^2);[*****]
- Operating force limits (release and opening) to 268 N (60 lbs); and
- Specific conditions for measuring opening forces, notably that emergency exit window shall be manually extendable by a single occupant to a position that provides an opening large enough to admit unobstructed passage, keeping a major axis horizontal at all times, of an ellipsoid generated by rotating about its minor axis an ellipse having a major axis of 50 cm (20 in) and a minor axis of 33 cm (13 in).

In addition, FMVSS 217 requires the following emergency exit operating force limits (release and opening):

- Motorcoaches: no higher than
    - 268 N (60 lbf) in in high-force regions and
    - 89 N (20 lbf) in low-force regions; and
- School buses: no higher than
    - 178 N (40 lbf) in high force regions, and
    - 89 N (20 lbf) in low-force regions.

[*****] This requirement ensures that manufacturers provide multiple emergency exits instead of just one large emergency exit on each side of the bus.

A drawing of an ellipsoid with a 50:33 proportion between its two major axes and its minor axis appears in Figure 6-1a. However, the actual test fixture used for exit opening compliance testing was built by rotating a 50 by 33 cm (20 by 13 in) ellipse around its major axis.

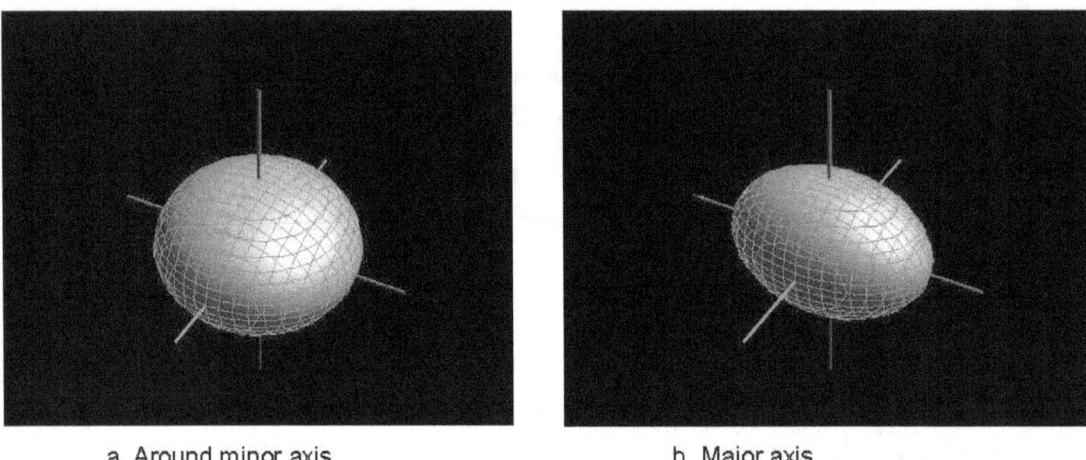

a. Around minor axis   b. Major axis

Figure 6-1.   **Ellipsoids Formed by Rotation of 50 by 33 cm (20 by 13 inch) Ellipse**

In response to a query from a bus manufacturer requesting clarification regarding the ellipsoidal test fixture, NHTSA has explained that its original intent was to require that conditions for opening force measurements be such that the ellipsoidal test fixture pass through the opening with the plane of its major axes horizontal, which would require a clear opening of at least 50 cm (20 in). However, the published regulation requires only "keeping a major axis horizontal at all times." Thus, tipping the ellipsoid, such that the clear opening is only 33 cm (13 in) between the window sash and the window sill, is permissible. Therefore, there is no practical difference between the two versions of the ellipsoid, i.e., both require an unobstructed opening 50 by 33 cm (20 in by 13 in).

The basis for these requirements derive from the 1970 OKRI motorcoach experiment[19] result that 2 out of 9 subjects could not fit through the 50 by 33 cm (18 by 13 in) ellipse (as reproduced in Figure 6-2), which defined exit minimum opening size as required by the then BMCS bus regulations.

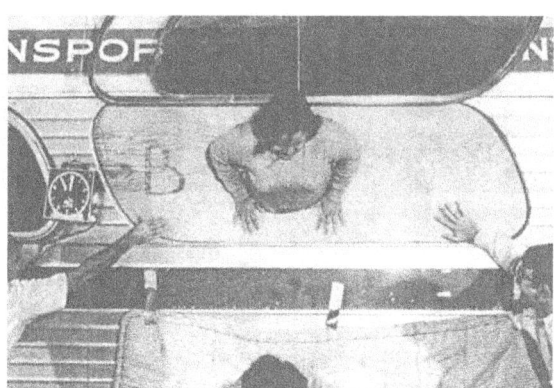

Figure 6-2. Subject Egress Using 50 by 33 cm (20 by 13 in) Elliptical Opening (OKRI 1970)

6.2 DESIGNS IN USE

6.2.1 Motorcoaches

To comply with FMVSS No. 217 requirements, a typical 55-seat motorcoach must have a minimum emergency exit area of 23,760 cm^2; which would be satisfied by six emergency exit windows, three on each side. The trend toward very large and heavy windows has made the emergency exit windows harder to use. The window weight – about 70 kg (150 lbs) for many recently built coaches – makes it difficult to raise them to a sufficient opening size. The large window surface area means that their retention mechanisms must be able to resist a force of 5,337 N (1,200 lbs), which complicates the design of an easily opened release. The need to seal out air and water leaks over a perimeter of more than 6 m (19 ft) places further demands on the window retention mechanism.

The conflict between the need for ease of opening and retention against large forces has prompted numerous design solutions. Even in motorcoaches of outwardly similar appearance and model number, there are differences in details of the retention mechanisms. Retention mechanisms in the national fleet installed by three of the major suppliers to the North American market are described in the next subsections.

6.2.1.1 *Motor Coach Industries (MCI)*

All of the *MCI* designs observed have a release bar at the sill that runs the full width of the window (Figure 6-3).

 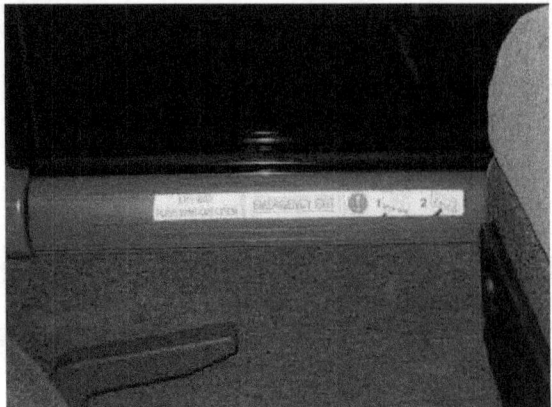

Figure 6-3.   *MCI* Emergency Exit Window Release Bars

More recent versions have a plastic cover over the aluminum extrusion (newer design on the right). To release the window, one must insert finger tips behind the bar from below and lift. There are twin latches near the sides of the window. The bar is somewhat flexible, so the use of both hands, spread as far as possible, provides the most effective application of force.

Many components of the retention mechanism are attached to the sill. The latches, normally concealed beneath the bar in the newer design, have been radically changed, as illustrated in the close-up views of Figure 6-4 (newer design on right).

Figure 6-4.   Close-Up View of *MCI* Sill-Mounted Emergency Exit Window Release Components

6.2.1.2 *Prevost*

Prevost (a Division of *Volvo*-Canada) has also used a long-bar release design in all of its coaches for the past two decades (see Figure 6-5). Like the newer *MCI* design, the lift bar is attached to the sash, which means that the bar protrudes into the open space between the emergency exit

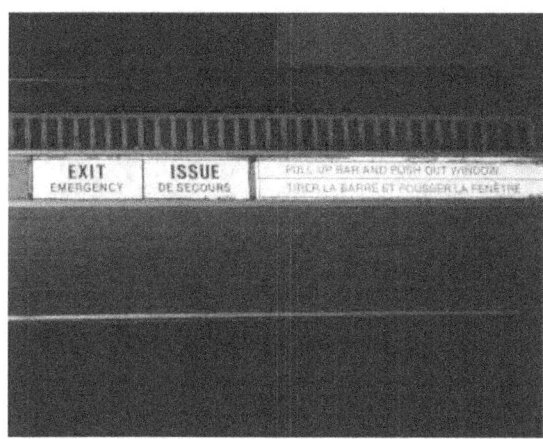

Figure 6-5. *Prevost* Sash-Mounted Emergency Exit Window Release Bars

window and the sill by about 7 cm (3 in). Thus, after release, these windows must be pushed further out by that amount to provide the same clear opening as the *Van Hool* window (see Subsection 6.2.1.3).

The *Prevost* window / latch design has also changed substantially (see Figure 6-6) (newer design on right).

Figure 6-6. *Prevost* Sill-Mounted Emergency Exit Window Latch Components

6.2.1.3 *Van Hool*

The *Van Hool* window release mechanism is entirely different (Figure 6-7) from the other types of exit windows, as it is contained almost entirely within the sash frame, with the result that the latch components on the sill are much smaller than those on *MCI* or *Prevost* windows. A red handle, located one side of each exit window, operates the release. The only evident changes between models are slight differences in the shape of the release handle and whether the handle is on the left or the right side of the window.

Figure 6-7. *Van Hool* Sash-Mounted Emergency Exit Window Release Mechanism

The small strikes located on the sill near each side of the *Van Hool* window (Figure 6-8), present much less interference with window egress than the sill-mounted mechanisms of the other manufacturers.

Figure 6-8. *Van Hool* Emergency Exit Window Sill-Mounted Release Components

6.2.2 School and Transit Buses

Two examples of school and transit bus emergency exit windows are shown in Figure 6-9. The red exit release handle for both windows operates the same way as the *Van Hool* motorcoach handle (see Figure 6-7). However, as Figure 6-9a shows, the school bus window is not hinged at the top, but is hinged to open on the side, when pushed open. These school and transit bus emergency windows exits are smaller in size than those installed in most motorcoaches.

        a. School bus   b. Transit bus

Figure 6-9.　　School and Transit Bus Side Emergency Window Exits -- Examples

6.3　ABILITY AND EGRESS RATES

Passenger egress through a bus emergency exit window involves a series of actions that can present significant physical challenges to some users:

- Releasing the window sash from its frame;
- Opening the window and keeping it in an opened position; and
- Traversing from the window sill to the ground.

The usability issues of the first two actions are discussed primarily in terms of the forces required to perform them, while results from human-subject experiment trials will be used to estimate the time required for traversing from the window sill to the ground, from which egress rates can be calculated.

6.3.1　Volpe Center Emergency Exit Window-Release Force Measurements

All measurements of emergency exit window–release forces were made with the *Omega Engineering LCCA-200*^{R} "S"-beam load cell, fitted with a hook on one side and a handle on the other, as illustrated in Figure 3-2.

Table 6-1 reports the highest (rounded) of the multiple measurements of forces required to release the emergency exit window from the frame for all of the U.S. motorcoaches tested.

Table 6-1. Motorcoach Emergency Exit Window – Peak Release Force Measurements

MAKE	MODEL	YEAR BUILT	WINDOW LOCATION CODE	PEAK RELEASE FORCE (N)	(lbf)
MCI	102DL3	1998	L1	62	14
	"	"	L3	298	67
	"	"	R3	119	27
	DL3	2000	R2	191	43
	J4500	2004	L1	143	32
	"	"	R1	107	24
	"	"	R3	89	20
	J4500	2006	L1	190	42
	"	"	R1	127	28
Prevost	LeMirage #120	1991	L1	124	28
	" #121	"	L2	152	34
	2000 #9300	2000	L1	66	15
	"	"	R2	131	29
Van Hool	C2045	1999	R1	73	16
	"	2001	L1	50	11
	"	"	R1	57	13
	"	2004	L2	134	30
	"	"	R3	165	35

All force measurements were well below the FMVSS 217 criteria limit of 268 N, except for one of the emergency exit windows on an older *MCI* coach tested in subfreezing conditions. The FMVSS 217 test procedure calls for testing at 70-85°F.) On this particular window exit, the release bar tended to flex considerably when force was applied at a single point near its center. A substantial portion of the force was dissipated in flexing the release bar, rather than opening the latches. Had the force been applied close to each latch, the values would have been much smaller. The most recent *Van Hool* bus appears to have higher release forces than earlier models.

6.3.2 Volpe Center Emergency Exit Window Opening Force Measurements

All of the motorcoach emergency exit window opening force values measured (see Table 6-2) are approximately consistent with those that could have been estimated from the calculation.

Table 6-2. Motorcoach Emergency Exit Window Opening Forces – Actual Measurements

MAKE	MODEL	SI UNITS			ENGLISH UNITS		
		Width (cm)	Opening# (cm)	Force (N)	Width (in)	Opening# (in)	Force (lbs)
MCI	102DL3	140	?	65	55	?	15
	"	"	?	53	"	?	12
	DL3	"		104	"	16.5	23
	J4500	165	25	91	65	10	20.5
	"	"	41	160	"	16	36
	"	"	56	196	"	22	44
	J4500	"	58	195	"	23	44
	"	"	"	201	"	23	45
Prevost	Mir. #120	84	43	74	42	17	17
	" #121	"	"	76	"	17	17
	2000 #9300	152	"	136	37	17	31
	"	"	"	131	"	17	29
Van Hool	C2045	165	46	110	56.5	18	25
	"	"	"	145	"	18	33
	"	"	43	128	65	17	29
	"	"	"	127	"	17	29
	"	"	66	260	77	26	58

"Opening" is the clear opening

6.3.3 Volpe Center Emergency Exit Window Experiment Egress Flow Rate Results

Figure 6-10 illustrates a subject during one of the two emergency exit window experiment small group trials described in Chapter 3.

The six subjects completed the egress in 42 seconds, equivalent to a flow rate of 9 ppm. During the debriefing, the subjects reported that they watched the person(s) immediately preceding and learned how to perform the window drop so that the rate of egress increased slightly as the trial proceeded.

The last subject (the +95th height-percentile male) elected to jump from the sill, as shown in Figure 6-11. This method made his egress the fastest observed. However, in jumping, he almost struck the window sash, despite the 91-cm (36-in) opening, as shown in the left-hand photo.

Figure 6-10. Subject Performing Motorcoach Emergency Exit Window Egress

a. Left – Jumping   b. Right – Dropping

Figure 6-11. Motorcoach Emergency Exit Window Egress – +95th Height Percentile Male

During the second set of trials, the +95th percentile subject was instructed to repeat the egress using the controlled drop procedure and staying as close to the bus as possible. Even with the preferable egress procedure, he needed a clear opening between the window sill and the sash of about 66 cm (26 in) for unobstructed passage through the window exit.

6.4 REQUIREMENTS FROM OTHER U.S. MODES

Passenger ship windows and emergency exit windows on passenger aircraft are so different from those on motorcoaches that their regulations have little application.

Passenger rail car emergency exit windows are roughly similar in size, weight, and height above ground to those in motorcoaches. Accordingly, FRA requirements in 49 CFR, Subsection 238.113,[30] are that at least four emergency window exits must be on each main level, distributed on both sides of the car, staggered, and placed near the ends of the car. Emergency exit windows on all new passenger rail cars must have required dimensions of 61 high by 66 cm wide (24 by 26 in). The interior location of these windows must be marked (see Chapter 8.)

FRA also requires "rapid and easy removal" by passengers of emergency exit windows from inside the passenger rail car. "Rapid and easy removal" has been interpreted to mean that the opening force may not exceed 50 pounds (222 N).

Finally, each new passenger car must have at least one emergency access point, either an exit hatch or a "soft spot," that is 61 in high by 66 cm wide (24 by 26 in). That location is required to be marked with retroreflective material on the exterior. Instructions for operation must also be provided on or adjacent to each roof access exit. (See Chapter 8.)

6.5 INTERNATIONAL BUS REQUIREMENTS

ECE 36[38] and ECE 107[39] include the following major provisions relating to bus emergency exit windows:

- Windows must open outward easily and instantaneously or be made of readily breakable safety glass.

- If hinged, windows shall be provided with an appropriate device to hold them fully open.

- Any windows not clearly visible from the driver's seat must be fitted with a device to provide an audible warning to the driver if the window is not latched.

- The test gauge for window openings is a thin plate 60 by 40 cm (24 by 16 in) with 20 cm (8 in) radius on the corners that must pass thru the opening while perpendicular to the direction a person would move in egress through the window.

There are no known quantitative definitions of "easily and instantaneously operated," but in practice, most European motorcoaches have fixed windows made of safety glass that is supposed to be broken using the hammer provided at each window. However, this type of glazing is far more likely to break in a rollover and lead to occupant ejection than the laminated-glass construction used by some North American bus manufacturers. European Commission members are considering an amendment that would require the use of laminated glass for windows.[61]

ADL 44/02 [41] is similar to the ECE regulations but it allows a third means of opening the window, i.e., a "window ejecting device." More significantly, it requires that windows that are more than 100 cm (39 in) above ground shall have a means to assist occupants in descending to the ground, such as footrests. The footrest may be a vehicle component.

ADL 58/00 [42] allows the use of windows as emergency exits, but does not specify performance requirements for such windows. It sets a maximum force requirement for removing the closing material for any type of emergency exit of 700 N (158 lbs) and a minimum force requirement of 445 N (100 lbs).

British Regulation No. 257 [43] does not explicitly require emergency exit windows, but apparently allows them to be included within the complement of secondary emergency exits. If used, window exits must provide a minimum aperture of 4,000 cm^2 (620 in^2) with minimum dimensions of 70 by 50 cm (28 by 20 in). Windows must either be "ejectable," hinged, or constructed of safety glass that can be readily broken by persons inside or outside the vehicle.

6.6 RELEVANT RESEARCH

In the 1970s, OKRI performed a series of motorcoach emergency egress experiments that included exiting through emergency exit windows.

During the 1970 OKRI egress experiments,[19] 10 subjects attempted to exit through a bus window fitted with plywood baffles containing elliptical openings of various sizes, as illustrated in Figure 6-2. All subjects were able to exit through the 61 by 43 cm (24 by 17 in) opening with a mean time of 2.47 seconds each. When the opening size was reduced to 50 by 33 cm (20 by 13 in), the average escape time increased to 3.96 seconds for the nine subjects who succeeded. One subject was injured and unable to complete this trial. For the third trial, the opening size was further reduced to 45 by 33 cm (18 by 13 in). The injured subject and a second subjects found it impossible to exit through an opening of this size (see also Table 2-1 in Subsection 2.1.4.1.)

Figure 6-12. 1972 OKRI Intercity Bus Egress Experiments (left and right side)

The OKRI authors concluded that a window opening size of 61 by 43 cm (24 by17 in) offers a significant advantage in egress rate over the smaller sizes.

During the 1972 OKRI study,[23] four egress experiments were conducted in which a group of 38 adult subjects exited from a GM PD4104 intercity bus by various means, as illustrated in Figure 6-12. Subjects were instructed to exit as quickly as possible, and in some trials, were allowed to use the front and rear doors, in addition to the windows.

During these four experiments, egress rates through a single emergency exit window as high as 13 ppm were noted in one case, but the typical rate was about 7 ppm per window. The emergency exit windows on this bus were much smaller than is typical of the current U.S. motorcoach fleet, and were held open by persons standing on the ground outside the coach.

In the 1978 OKRI experiments,[24] 45 subjects were used along with a *GM PD-4107* coach. Trials 1 and 2 were conducted under conditions of simulated darkness by using goggles. Subjects escaped from the upright coach through window exits at rates of 1 to 23 ppm, after taking from 3 to 33 seconds to open the push-out windows.

A more recent study performed in 1984 by the Hungarian bus manufacturer, *IKARUS*, used a group of 45 male firefighters, aged 20 to 40 years, as subjects in a window-evacuation test.[56] They were instructed to kick out all of the windows on both sides and exit as rapidly as possible, which they accomplished in only 10 seconds; equivalent to an egress rate of 34 ppm per window (see Figure 6-13a).

a. Firefighters   b. Civilian female

Figure 6-13. 1984 Hungarian Bus Window Egress Tests

The IKARUS study also tested a 30-year-old female subject who was unable to shatter the window with the first hammer provided, and took 15 seconds to complete the task when given a better tool. Clearing most of the glass consumed an additional 25 seconds. She was described as, "afraid of jumping through the window, which had glass fragments in the "waistrail" (i.e., windowsill)." Thus, she needed help from outside to jump out, as illustrated in Figure 6-13b and required an additional 50 seconds to complete her egress, with assistance by the firefighters outside the bus.

A recent study conducted for the British (UK) Department of Trade and Industry[47] raises questions about the 268-N force limits currently permitted by FMVSS 217 criteria for releasing and opening any of the means of egress on a motorcoach. Subjects in this UK study were tested to determine the maximum force they could exert on the apparatus shown in Figure 6-14).

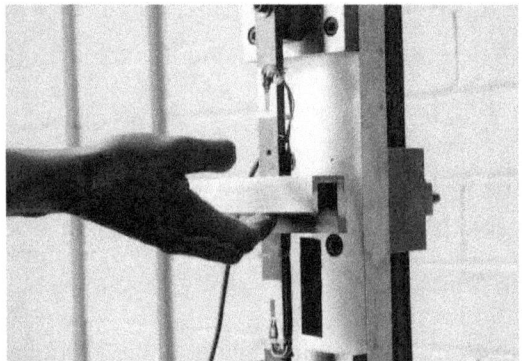

Figure 6-14. Test Apparatus for Measuring Finger-Tip Pulling Strength

This UK strength-measurement test is similar to the task that must be performed to pull the release bar on the majority of bus emergency exit windows. The distribution of maximum forces exerted by 152 subjects is shown in Figure 6-15. The data in Figure 6-15 show that that most of the subjects in these tests were not able to exert forces of 268 Newtons, which is the current allowable FMVSS 217 maximum limit.

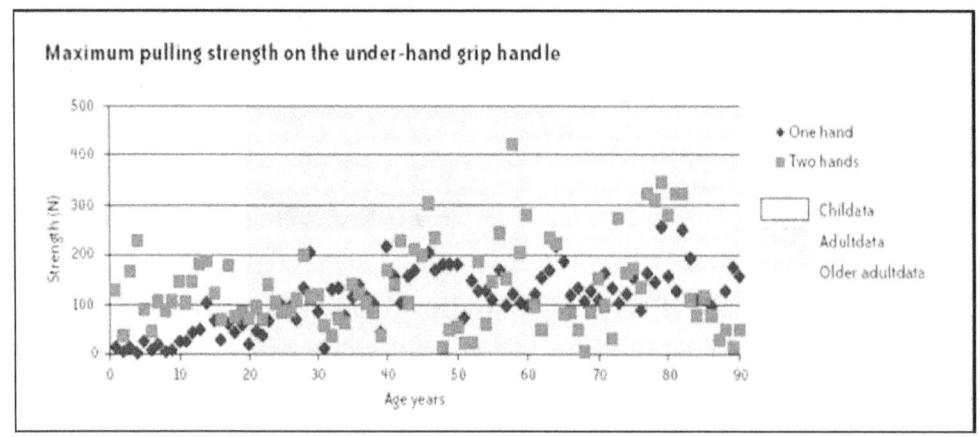

Figure 6-15. UK Subject Distribution – Maximum Finger-Tip Pulling Strength

6.7 DISCUSSION

In crashes in which the motorcoach remains upright but the front door is blocked or jammed, windows are the principal means of emergency egress in the current fleet of motorcoaches. Experiments conducted by OKRI and the Volpe Center indicate that a fully loaded motorcoach could be evacuated via the emergency exit windows in less than two minutes, provided that the occupants have the requisite strength to open the exits, that means of holding the windows open are available and that passengers have enough agility to move through the exits.

The FRA dimension requirement for new passenger rail cars of an unobstructed emergency window exit and access opening of at least 61 in high by 66 cm wide (24 by 26 in) is based on MIL-STD 1472F[45] performance criteria, as well as the consideration that emergency responder evacuation stretchers are approximately 66 cm (24 in) wide.

There has been a trend toward larger and heavier windows in newer motorcoaches, which has made egress through them substantially more difficult. FMVSS 217 requires "hold open" devices for school bus emergency exit doors. However, there are no "hold-open" devices required or currently provided for U.S. motorcoach emergency exit windows. The design of a "hold-open" mechanism requires a tradeoff between exit opening size and opening force limits.

In its 1999 bus crashworthiness issue report,[11] NTSB expressed concern about how that trend toward larger windows and fewer pillars supporting the roof degrades crashworthiness, in terms of both structural integrity and occupant ejection risk.[9] Design changes by motorcoach manufacturers to address these concerns – essentially making emergency exit windows smaller and lighter – would also facilitate emergency egress by passengers. Emergency exit windows that are not hinged at the top, i.e., side-hinged or sliding which FMVSS 217 has permitted since May 1995, could also be installed in motorcoaches to eliminate the need to overcome the force of gravity in opening large windows.

Although ECE requires that all hinged windows must be provided with an appropriate mechanism to hold them open, Volpe Center staff did not identify any buses in use as having those "hold-open" devices installed.

Although FMVSS 217 requires positive "hold-open" devices for school bus emergency exit doors, motorcoach window installation and operation during an emergency are significantly different, posing a design challenge to bus manufacturers.

It is not possible to predict what motorcoach egress flow rates could be achieved if the emergency exit windows were not being held open by some means. Attempts to exit through

68 kg (150-lbs) windows without "hold-open" devices are likely to cause injury, and experiments to test this mode of egress are unlikely to be approved by an IRB. Several subjects were injured by emergency windows exits during the OKRI studies, even though those bus windows were much lighter than those of the current fleet and the window height above the ground was such that persons standing on the ground could hold them open, as previously shown in Figure 6-12.

(Note: This scope of this study does not address the issue of window retention - including emergency exit windows - during motorcoach crash impacts.)

With the exception of one sample tested in freezing conditions, all of the motorcoach emergency exit windows tested conformed to current FMVSS 217 release-force limits (268 N).

All tested samples were within the 268-N limit for opening forces, for openings as large as 66 cm (26 in); however, one of the heavier windows measured 260 N at that opening size. In addition, the recent UK human-strength research results[47] imply that a large portion of the general population may have difficulty in providing the required force to open the window release bars used on many motorcoaches. Accordingly, further study is needed to develop appropriate force limits.

6.8 CONSIDERATIONS

Potential motorcoach design changes that may increase the passenger egress rate and reduce the risk of passenger injuries during emergencies include:

- Use of "hold open" retention mechanisms to keep emergency exit windows in an open position, sufficiently large to permit unobstructed passage of 95th percentile male subjects for large buses other than school buses – an opening of at least 61 cm (24 in) high;
- Clearer language specifying how to perform the required emergency exit window-opening-force measurements; and
- Instructional signage and pictograms explaining how to use the emergency exit windows should show how to prop windows open and how to perform a controlled drop.

The existing research literature is inconclusive with regard to the force levels that can be applied to the emergency exit window-opening tasks by the general public. Volpe Center staff are conducting human-subjects tests during the second year of this study to measure these human strength aspects, in a mock-up simulator that exactly reproduces the geometry of an actual motorcoach.

7. EMERGENCY ROOF EXIT EGRESS

The NHTSA 2007 research plan stated that 49 percent of motorcoach fatalities were associated with a rollover as the most harmful event.[3] When a bus is lying on its side, an emergency roof exit hatch is the only readily accessible emergency exit available for use, and is much less hazardous than climbing up to either the emergency exit windows, or the door (if not blocked), on the upward side of the motorcoach, both of which could be approximately 244 cm (8 ft) above the ground. Moreover, the NTSB recommendation that other than floor-level emergency exits be able to be easily opened and remain open during an emergency evacuation when a motorcoach is upright or at unusual attitudes, e.g., overturned (H-99-9),[8] applies to emergency roof exit hatches.

7.1 FMVSS 217 REQUIREMENTS

FMVSS 217 requires that motorcoaches be equipped with:

- Rear emergency exit doors; but emergency roof exit hatch(es) may be substituted if a rear emergency exit door is not provided;

- Roof exit hatches with operating forces of no more that 268 N (60 lbs); and

- Roof exit hatches that provide unobstructed passage of the same 50 by 33 cm (20 by 13 in) ellipsoidal test fixture described in Section 6.1.

FMVSS 217 requires that school bus emergency roof exits provide an opening when the exit hatch is extended that has minimum dimensions of 41 by 41 cm (16 by 16 in). In addition, school bus roof hatches are required to be hinged on their forward side and be operable from both inside and outside. A school bus with two roof hatches must have them placed at approximately one-third and two-thirds of the length of the bus. A school bus with three roof hatches must have them placed at approximately one-fourth, one-half, and three-fourths of the length of the bus. FMVSS 217 states that all school bus emergency roof exit hatches should be installed near the centerline of the vehicle, or if this is not feasible, there should be equivalent offsets on both sides of the centerline.

7.2 DESIGNS IN USE

Every motorcoach in revenue service encountered during this study had two emergency roof exit hatches, with minimum dimensions larger than those currently required by FMVSS 217.

There is much less variety in emergency roof exit hatch design than is the case for emergency exit windows. All of the *MCI* motorcoaches observed were fitted with oblong exit hatches (measuring about 87 by 52 cm (20.5 by 34.5 in), as shown in Figure 7-1. The hatch mechanism has changed over time from the stainless steel handle shown on the left to the red plastic one on the right. However, the required action is the same – pull the handle down as far as possible to release and push the roof hatch open with the other hand. Moving the handle to this position and the addition of the pictogram likely helps to guide users to apply the opening force more effectively, as far from the hinge as possible. These emergency roof exit hatches weigh about 18 kg (40 lbs), which means they are somewhat difficult to open when the motorcoach is upright. However, when the motorcoach is on its side, the hatches swing open freely.

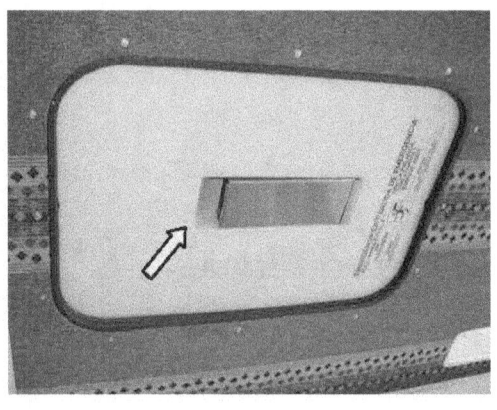 

a. Older   b. Newer

Figure 7-1. *MCI* **Emergency Roof Exit Hatches – Interior**

Prevost and *Van Hool* motorcoaches use emergency roof exit hatches that are about 56 by 56 cm (22 by 22 in) – apparently both from the same supplier (see Figure 7-2).

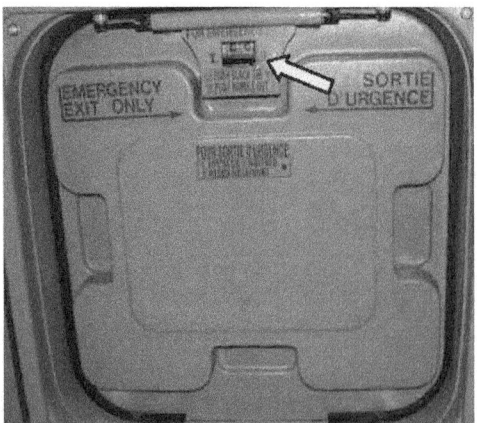

Figure 7-2. *Prevost* **and** *Van Hool* **Emergency Roof Exit Hatch**

Transit buses also use roof hatches designed similar to that of Figure 7-2. The required action to release and open the hatch is to push the black tab in the direction indicated by the arrow and then push out ward on the hatch. These roof exit hatch covers weigh only about 9 kg (20 lbs), and, like the ones used by *MCI*, swing freely when the motorcoach is on its side (Figure 7-3).

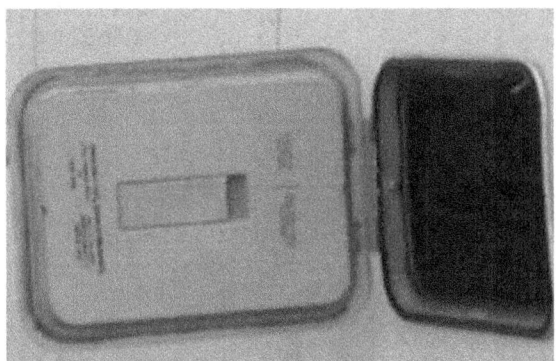

Figure 7-3. *MCI* Open Emergency Roof Exit Hatch – Exterior

Similar emergency roof exit hatches are used on school buses; however, those hatches are required to be marked on the exterior with "Emergency Exit" signs and have an exterior release handle. Figure 7-4 shows a typical emergency roof exit hatch from the interior and the exterior of the school bus.

 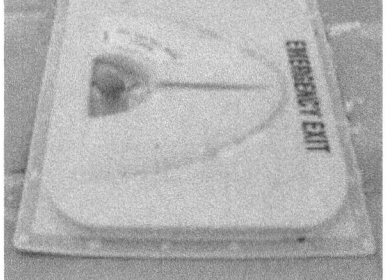 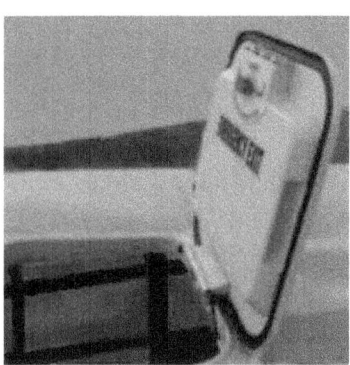

a. Interior   b. Exterior – closed   c. Exterior – open

Figure 7-4. School Bus Emergency Roof Exit Hatch

All emergency roof exit hatches include features to allow them to be held open a few inches while the bus is upright, so that they can provide ventilation in the event of a failure of the air-conditioning system. However, none contain any provision to hold them fully open at all times. When the motorcoach is upright, gravity will keep the hatch open once it has been pushed past the vertical, but when the coach is on its side, the cover could swing in the wind and can pinch the hands of individuals attempting to escape through it. All roof exit hatches tested by Volpe

Center staff released with relatively modest amounts of force, as indicated in Table 7-1 (SI and English units).

Table 7-1. Volpe Center Emergency Roof Exit Hatch Release Force Measurements

MAKE	MODEL	DIMENSIONS cm (in)	RELEASE FORCE N (lbs)	UPRIGHT OPENING FORCE N (lbs)
MCI	D	87 x 52 (34.5 x 20.5)	27-62 (6-14)	49-76 (11-17)
	J	" (")	45-62 (10-14)	133-156 (30-35)[#]
	"		27-36 (6-8)	133 (30)[#]
Prevost	2000	56 x 56 (22 x 22)	67-90 (15-20)	175 (39)[##]
Van Hool	C2045	" (")	48 (11)	37 (8)
	"	" (")	28 (6)	21 (5)
	"	" (")	76-93 (17-21)	30 (7)

[#] Force not applied at most effective location, i.e., force should have been applied as far from hinge as possible.

[##] Likely an artifact caused by gauge impacting stop when hatch popped up to "ventilate" position.

7.3 USABILITY AND EGRESS RATES

During the 2008 roll-over tests conducted at the MGA test facility, the emergency roof exit hatches on both motorcoaches flew open as the vehicles impacted the pavement.[62] Afterwards, all exit hatches latched and reopened normally. Hatch opening forces were not measured with a gauge, but were estimated to be only a few Newtons, and therefore well within required FMVSS 217 limits. The principal issues were that the emergency roof exit hatches did not stay open during the Volpe Center research team egress tests and tended to slam shut in wind gusts.

All of the emergency roof exit hatches tested were well below FMVSS 217 required release force limits. When motorcoaches were on their sides, hatch-opening forces were negligible. With the motorcoach upright, the force required to raise the emergency roof exit hatch is equal to one-half of its weight, so long as that force is applied as far from the hinge as possible.

However, roof exit usability and egress rate are strongly affected by other factors which include:

- Angular orientation of the motorcoach body;
- Dimensions of the hatch (es);
- Number of hatches;
- Illumination; Presence of hand and footholds (interior and exterior) to facilitate use of the hatch, and
- Strength and agility of users.

As indicated in Chapter 3, Volpe Center staff members conducted a limited experiment using the two types of emergency roof exit hatches after the two motorcoaches had been tipped over on their sides at the MGA test facility. The *MCI* bus roof exit hatch opening was 87 by 52 cm (34 by 21 in), while the *Prevost* motorcoach hatch opening was 56 by 56 cm (22 by 22 in).

The three methods of exit through an emergency roof exit used during the limited egress experiment and their estimated flow rates when a motorcoach is on its side were:

- Somersault egress - about 4 seconds, which implies a flow rate of 15 passengers per minute.
- Whole-body lift - about 5 seconds, implying a flow rate of 12 passengers per minute.
- Cautious approach - about 40 seconds for the *MCI* motorcoach and well over one minute the *Prevost* motorcoach, with the slightly smaller hatch opening.

In a time-critical evacuation scenario, it would be advisable for passengers to assist each other, out of the emergency roof exit hatch, which should provide a much more rapid flow than the cautious approach.

7.4 REQUIREMENTS FROM OTHER U.S. MODES

Emergency roof exits are not generally required in transportation modes, although some U.S. railroads have installed them on some locomotives. FRA does require that new U.S. passenger rail cars have either two marked roof hatches, or "soft spots" in their roofs, which emergency responders can open or cut through to gain access for passenger rescue.[30] The soft-spot option is prepared by both railroad and manufacturer preference due to design and safety concerns.

7.5 INTERNATIONAL BUS REQUIREMENTS

ECE 36 [38] and 107 [39] require the following of all motorcoach escape exits:

- At least two exit hatches, if the vehicle seats over 50 passengers;

- A minimum hatch aperture area of 4,000 sq cm (620 sq. in), with a minimum dimension of 50 cm in one axis and 70 cm in the other axis (20 by 28 in);

- Capable of being easily opened from inside or outside; and

- Interlocks and warning devices to prevent unintentional opening (for floor hatches).

ADR 44/02[41] does not require emergency "escape" hatches, but allows such hatches to be used to meet requirements for the total number of emergency exits. If used, they must have apertures of at least 4,000 cm^2 (620 in^2), with minimum dimensions of 50 cm by 60 cm (20 by 24 in). Both sliding panels and erectable (hinged) roof hatches are permitted, and sliding panels are allowed to have opening forces of as high as 500 N (113 lbs).

ADR 58/00[42] requires either a rear door or a roof hatch in combination with a side door on the opposite side from the service door. The minimum aperture is 3,200 cm^2 (496 in^2).

British Regulation No. 257[43] requires an emergency roof exit hatch, unless a bus is fitted with an emergency exit in the front or rear of the vehicle. The exit area and dimension requirements are the same as in ECE 36.

7.6 RELEVANT RESEARCH

Conducting realistic emergency roof exit hatch-egress experiments could be difficult and expensive because of the need to either turn an actual motorcoach on its side or build a "mock-up" for occupant egress simulation purposes. The 1972 and 1978 OKRI studies provide the only known realistic U.S. bus emergency roof exit hatch experiments.[23,24]

In the 1972 experiments, plywood mock-ups of school bus emergency roof exit hatches were used as illustrated in Figure 7-5. The openings were 61 by 61 cm (24 by 24 in) in the left photo and 61 by 102 cm (24 by 40 in) in the right one.

The children were able to crawl through at rates of about 30 per minute per hatch for both configurations. However, the 1972 trial results have little applicability to real roof hatches in motorcoaches or school buses because use requires that passengers somehow raise themselves up about 1.2 m (4 ft) from the surface location on which they start, which is challenging and time consuming for many persons.

As described in the 1978 intercity bus egress report,[24] OKRI tipped a *GM PD 4107* motorcoach on its side and tested egress rates through an actual roof exit hatch with adult subjects.

Figure 7-5.   School Children Crawling through Roof Hatch Mockups (OKRI 1972)

Two trials were held with goggles that simulated emergency-lighting conditions, i.e., an average perceived luminance of 0.2 foot-lamberts (685 mcd/m^2). Two other trials were conducted with subjects wearing goggles that simulated darkness (average perceived luminance of 0.005 foot-lamberts (17mcd/m^2).

In the first trial conducted in darkness, 12 subjects exited via the roof exit hatch at an average rate of 6.64 passengers per minute, while the remainder of the 45 subjects used the window emergency exits. Use of the windows was facilitated by a platform with steps that was erected alongside the exterior of the overturned coach. The egress rate in the second darkness trial was essentially identical – 6.7 ppm for 11 subjects. For reasons not entirely clear, the egress rate dropped to 4.64 ppm in the first trial that was held with goggles that simulated emergency lighting in which 7 subjects used the roof hatch. However, in the final trial held in the "emergency-lighting" condition, the egress rate rose to 10.67 ppm, again with 7 subjects. However, none of the differences were statistically significant given the very small number of subjects involved.

7.7   DISCUSSION

More than half of the fatal motorcoach crashes involve rollovers. When a motorcoach is on its side, it is extremely difficult or impossible to use the front door or emergency exit windows. For this reason, all U.S. motorcoaches are required to have at least one emergency roof exit hatch, if they are not equipped with a rear emergency exit door. Every motorcoach in revenue service observed by Volpe Center staff during the bus company operator and the MGA test field visits has been equipped with two emergency roof exits.

The FRA new passenger rail car dimension requirement for an unobstructed emergency roof access opening of at least 61 cm high by 66 cm wide (24 by 26 in), is based on MIL-STD 1472[45] performance criteria, as well as the consideration that the great majority of emergency responder evacuation stretchers are approximately 66 cm (24 in) wide.

As noted previously, bus emergency roof exit hatches may not stay open when fully released. The equipping of emergency roof exit hatches with positive "hold open" devices would address this issue and the NTSB recommendation that other than floor-level door emergency exits remain open when a motorcoach is overturned.[10]

It is unlikely that a full load of passengers could be evacuated through motorcoach roof hatches of current design within the time available in a fire or water immersion scenario. Design issues relating to motorcoach emergency roof exits include:

- What provisions are made to help the passenger user climb through the roof hatch?

- How large is the roof hatch? (Bigger is better in terms of egress rate, and all hatches in use in U.S. motorcoaches are larger than FMVSS 217 requires.)

- Is there sufficient illumination to read the instructions and see how to use the roof hatch?

- How is the roof hatch secured? (Current designs are easy to open, but may not prevent occupant ejection in rollovers.

- Provisions for opening the roof hatch from outside by emergency responders. This is currently required for school buses, but not for motorcoaches.

7.8 CONSIDERATIONS

Potential motorcoach design changes that may increase the passenger egress rate and reduce the risk of injuries during emergencies, particularly when the motorcoach is turned on its side, include:

- Minimum emergency roof exit hatch aperture dimension of 4,000 cm^2 (620 in^2);

- At least two emergency roof exit hatches installed on each bus. (Many bus manufacturers of motorcoaches operated in the U.S. have already adopted two hatches as standard practice for buses.);

- Handholds/footholds on the interior of the roof, or luggage bins designed to incorporate equivalent supports for passengers using the hatch;

- "Hold-open" devices to hold emergency roof exit hatch covers open;

- Instructional pictograms on the interior of roof hatches illustrating how to use hatches to exit the overturned bus; and

- Exterior operation capability of roof hatches and retroreflective marking and instructions for emergency responder access, consistent with current FMVSS 217 school bus requirements.

8. EMERGENCY EXIT IDENTIFICATION

Other common carriers of passengers, are required to locate "Emergency Exit" signs on or above all doors, windows, roof (and floor) hatches, etc., that may be used for emergency egress. In addition, "exit locator" emergency signage (to guide passengers to the emergency exit) and "low-location" exit path marking is often required to assist passengers in locating and moving towards exit crawling under smoke conditions. NTSB has recommended that NHTSA require that motorcoaches use improved conspicuous emergency signage, so that when necessary, passengers will be able to clearly identify, reach, and operate the emergency exits more quickly (H-00-02).[10] The 2007 NHTSA research plan identified the feasibility and desirability of having emergency signage consistent with other public transportation as a priority consideration[3].

8.1 FMVSS 217 REQUIREMENTS

FMVSS 217 motorcoach emergency exit identification requirements specify that the interior of each designated emergency exit door must be identified with "Emergency Door" or "Emergency Exit." All other designated motorcoach emergency exits must be marked with "Emergency Exit" and concise operating instructions describing each motion necessary to unlatch and open the exit, which must be located within 16 cm (6 in) of the release mechanism. Examples of instructional text are:

1. Lift to Unlatch, Push to Open
2. Lift Handle and Push out to Open.

When an emergency exit release mechanism is not located within an occupant space of an adjacent seat, a label meeting legibility requirements that indicates the location of the nearest release mechanism must be placed within the occupant space. For example: "Emergency Exit Instructions Located Next to Seat Ahead."

FMVSS 217 does not include explicit requirements for motorcoach emergency exit signage, including window or roof exit hatch instructions such as marking letter size or type of material. However, these instructions must be legible to occupants with corrected visual acuity of 20/40 (Snellen ratio) under normal nighttime illumination.

As noted in Chapter 9, FMVSS 217 requirements are not specific as to what "normal nighttime illumination" means. This performance requirement is easily satisfied by all of the installed emergency exit signage observed by Volpe Center staff when the bright fluorescent lights normally used during loading and unloading were in use. However, none of the emergency exit signage was legible when only the nightlights (see description in Section 9.2) were operating. Moreover, sign legibility is variable and indeterminate when some proportion of the reading lights are in use.

FMVSS 217 school bus emergency exit signage and instruction requirements differ from those for motorcoaches in that there are explicit requirements that all emergency exits be marked with signs using letter heights of 5 cm (2 in) and with instructions with a letter height at least 1 cm (3/8 in), both of which are required to be a color that contrasts with its background. The emergency sign designation marking must be located at the top of or directly above the emergency exit door, on the inside (and outside) of the exit. For emergency window exits, the signage must be located at the top of, directly above, or at the bottom of the emergency window exit on both the inside and outside surfaces.

The signage for school bus emergency exit hatches must be located on the inside of the bus within 30 cm (12 in) of the roof exit opening.

Concise instructions for operating all emergency exit including unlatching and opening all emergency exits must be located on the inside of the school bus within 15 cm (6 in) of the operating mechanism. The minimum letter size for instructions must be 1 cm (3/8 in) and of a contrasting color with the background.

In addition, the perimeter of all school bus emergency exits are required to be marked on the exterior with a 2.5-cm (1-in)-wide outline of red, white, or yellow retroreflective material that meets minimum retroreflectivity criteria specified in Table 1 of FMVSS 217.

8.2   DESIGNS IN USE

As noted in Chapter 4, in all other transportation modes, "service doors" (the ones through which passengers normally enter and exit) are also regarded as "emergency exits" for passenger egress use, if necessary. As such, those door exits are always marked wit signs (including "locator" (see FAA requirements in Section 8.3) signs to identify exit location if the exits are hidden by a bulkhead or other obstruction), as well as instructions for opening them.

However, FMVSS 217 requires manufacturers to determine which exits are considered emergency exits necessary to meet the bus and school bus emergency exit requirements. Some service doors are marked as emergency exits, but most of those observed by Volpe Center staff during the opening-force tests were not.

Because front doors are normally power actuated, some motorcoaches contain signage intended to explain how to open the door in the event that the driver is incapacitated or that the power-actuation system malfunctions.

Figure 8-1 shows an example of motorcoach front-door operation instructional interior signage.

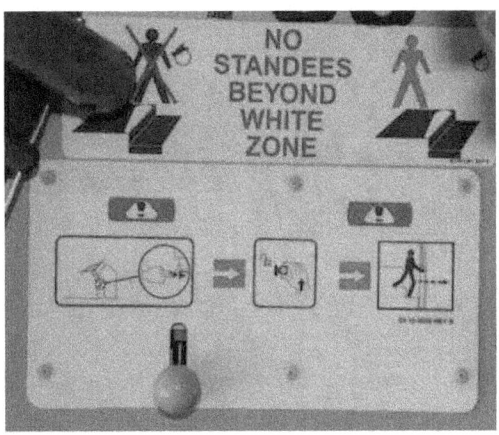

Figure 8-1.   Motorcoach Front-Door Operation Instructional Signage Example – Interior

This sign is located above the door of the motorcoach, where it is out of view of a person standing in the step well and attempting to open the door. Moreover, during field testing and setup for the 2008 Volpe Center-conducted human-subject experiments, various staff members had initial difficulty in interpreting its meaning, and sometimes took more than one minute to determine how to open the door. (The knob is not red in color as most emergency exit release controls are.)

Figure 8-2 shows an example of the clearer interior school bus front-door-opening emergency exit instructions, located next to the red switch at the right top of the door.

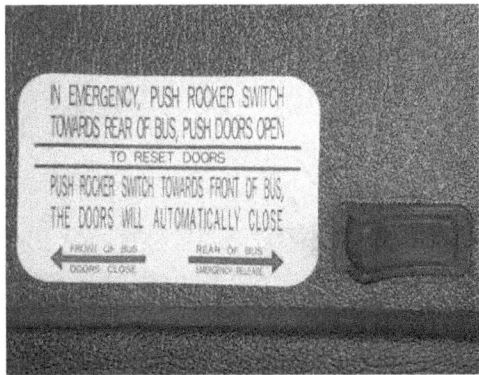

Figure 8-2.   School Bus Front-Door Instructional Signage Example – Interior

All of the motorcoach emergency window and roof-hatch exits observed by Volpe Center staff use 1 cm (3/8 in) letter heights for the emergency exit signage and instructions at both locations.

Although current motorcoach emergency exit window signage and instructions vary in their conspicuity,****** they are clearly visible to passengers seated next to the exit during normal daytime illumination (see Figure 8-3 and Figure 8-4.).

Figure 8-3. Motorcoach Emergency Exit Window Marking and Instructions (1)

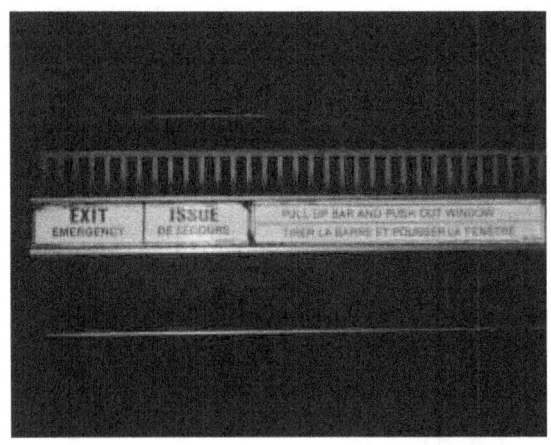 

Figure 8-4. Motorcoach Emergency Exit Window Marking and Instructions (2)

In addition, *Van Hool* motorcoach emergency window exits observed by Volpe Center staff had emergency exit signs, comprised of red letters, installed on the window surface (see Figure 8-5). However, the larger sign located on the window, just under the luggage rack, in Figure 8-5b, is not required by FMVSS 217.

****** "Conspicuity" is defined as: visual characteristic that is highly recognizable and attracts attention, by using size, brightness, and high contrast.

 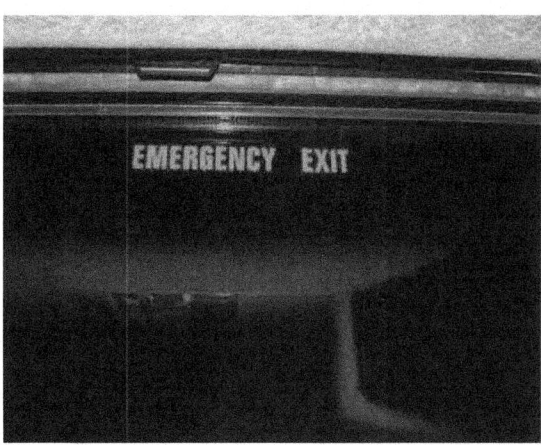

a. Marking and Instructions   b. Sign

Figure 8-5. Motorcoach Emergency Exit Window Signage and Instructions

Figure 8-6 shows school bus side window emergency exit marking that complies with FMVSS 217 requirements for signs and instructions.

a. Interior   b. Exterior

Figure 8-6. School Bus Window Emergency Exit Signage

Figure 8-7 illustrates the interior signage of transit bus doors and emergency window exits, as well as operating instructions.

Figure 8-7. Transit Bus Door and Window Emergency Exit Signage – Interior

Motorcoach and school bus interior emergency roof exit emergency signage and instructions are shown in Figure 8-8. Emergency roof exit emergency signage and instructions can be unclear as shown in Figure 8-8a, which illustrates how a person might use the exit hatch to escape from an upright motorcoach. However, the hatch is needed primarily for egress from overturned coaches. Figure 8-8b has pictograms on the lower part of the roof exit hatch showing the bus on its side and how to use the hatch in that condition. Figure 8-8c shows the school bus roof hatch marked with the 5 cm-high (2 in) letters, as required by FMVSS 217.

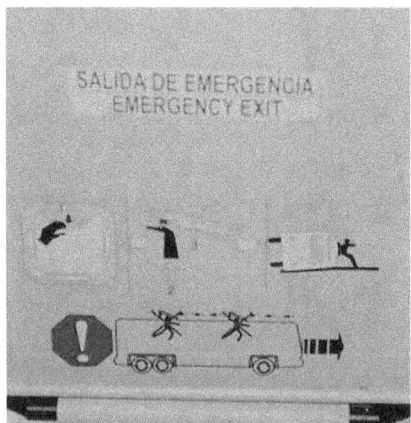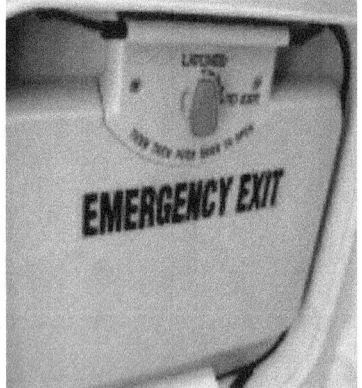

a. *MCI* – older   b. *MCI* – newer   c. School bus

Figure 8-8. Motorcoach and School Bus Emergency Roof Exit Hatch – Interior Signage and Instructions

Figure 8-9 shows examples of retroreflective marking used to designate school bus emergency door, window, and roof hatch marking (reflecting camera light flash).

  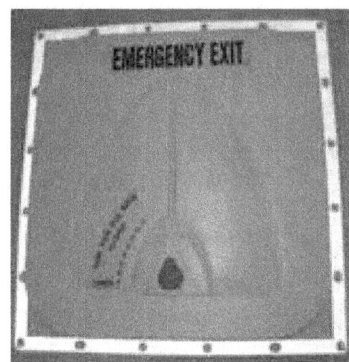

a. Rear door exit b. Side window exit c. Roof hatch exit

Figure 8-9. School Bus Emergency Exit – Exterior Retroreflective Marking

8.3 REQUIREMENTS FROM OTHER MODES

8.3.1 Interior Identification

All other transportation modes require interior emergency signs and instructions for all emergency exits. In other modes, normal service doors are also considered to be emergency exits. Emergency sign letter height requirements vary with the size of the vehicle and the expected distance from the viewer to the designated sign.

Principal requirements include:

- *Aviation* – 14 CFR, Subsection 25.811 and 812[29] – FAA requires that each interior emergency exit, its means of access, and means of opening must conspicuously marked. The identity and location of each exit must be recognizable from a distance equal to the width of the cabin.

  Each emergency exit must be identified by electrically-powered emergency exit signs next to each emergency exit, as well as "locator signs," ******* both of which can be located above the aisle near each emergency exit or at another overhead location, if more practical. One sign can serve more than one exit if each exit can be seen readily from the sign.

  The location of the operating handle and instructions for opening certain types of door exits from the inside must be shown by a marking on or near the exit that is visible from at least 76 cm (30 in).

******* This sign identifies the exit location fore and aft in the cabin if the exit is beyond and obscured by the bulkhead.

The emergency lighting requirements for these signs vary with the seating capacity of the aircraft and the location of the sign. For aircraft with seating for 10 or more passengers, the overhead exit locator signs must have a minimum background luminance of 25 foot-lamberts (86 cd/m^2), while the bulkhead signs must meet a minimum of 400 microlamberts (1.27 cd/m^2). For aircraft with 9 or fewer passengers, a lower initial background luminance of only 160 microlamberts (0.51 cd/m^2) is required for their emergency exit signs.

Additional FAA emergency exit sign requirements, including emergency lighting for those signs, are described in Chapter 9.

- *Rail* – 49 CFR Subsections 238.235 and 238.113[30] – FRA requires that passenger rail car emergency signs for doors used as emergency egress and emergency exit windows be "conspicuous" and marked with "luminescent" material (i.e., that absorbs light energy when light levels are high and emits the light when levels are low and appears to glow in the dark). Figure 8-10 shows examples of such signage.

Clear and legible instructions for operating the exit doors and windows must be located at or near each exit.

Figure 8-10. Passenger Rail Car – Photoluminescent Emergency Exit Signage

- *Marine* – 46 CFR, Subsection 122.606[35] – USCG specifies 5-cm (2-in) high letters for required emergency exit signage installed on passenger ships.

The APTA industry standard for passenger rail car emergency signage specifies 3.8 cm (1.5 in)-high letters for vestibule and side exit doors and 2.5 cm (1 in) letters for emergency exit windows and requires that "high performance photoluminescent" (HPPL) material[********] be used for door emergency exit signs.[33]

[********] "High-Performance Photoluminescent" material exhibits significantly enhanced surface brightness for a much longer time period compared with zinc sulfide photoluminescent material.

FRA has indicated that it plans to incorporate the APTA emergency signage standard by reference into the passenger rail regulations in a future rulemaking.[31]

FAA and IMO / SOLAS[37] also require that passenger aircraft and passenger ships have floor proximity / "low-level" exit-path-marking systems, respectively, to mark the interior emergency exit locations and the floor aisle escape route, so these locations remain visible to occupants when the primary emergency exit signs are obscured by smoke.

IMO / SOLAS[37] allows the passenger ship exit path marking system to be implemented using either photoluminescent marking or using "low-level" lighting using lights with self-contained power sources located near the floor. If PL materials are used, they must meet a criteria of at least 15 mcd/m^2, measured 10 min after the removal of all external illuminating sources and provide luminance values greater than 2 mcd/m^2 for 60 min.

FAA has issued two advisory circular (AC) guidance documents that provide information to assist aircraft manufacturers and airline operators in demonstrating compliance with the exit path requirements, by the means of observation tests.[63][64] These guidance documents are discussed in Chapter 9.

The APTA passenger rail car "low-location" path marking industry standard[34] also requires that either "high-performance photoluminescent" markings or a type of independently powered electrical lighting system similar to that required by FAA and IMO /SOLAS be used.

8.3.2 Exterior Identification

FAA and FRA both require that the exterior of aircraft and passenger rail car emergency exits and rescue access locations be conspicuously marked with reflective material.

Principal requirements are:

- *Aviation* – 14 CFR, Subsection 25.812[29] – FAA requires that each passenger emergency exit that is required to be openable from the outside, its means of access, and its means of opening must be conspicuously marked on the outside of the aircraft.

    The exit exterior marking must include a 5 cm (2 in) band of color outlining the exit, which must have a high color contrast to be readily distinguishable from the fuselage.

    For other than side door exits, the outside marking must be a red or bright chrome yellow if the red color is inconspicuous.

- *Rail* – 49 CFR, Subsection 238.115[30] – FRA requires the use of retroreflective markings to conspicuously identify the exterior location of passenger rail car doors, windows, and roof hatches or "soft spots" that emergency responders can use to gain access to assist passengers in evacuating a train.

All retroreflective marking material must be of high color contrast to the surrounding surface of the rail car.

A unique and easily recognizable symbol retroreflective marking is required for all designated rescue access doors and windows. The roof "soft spot" must be outlined by a border of retroreflective material that is 2.5 cm (1 in) in width.

FRA requires the use of "Type I" retroreflective material for exit exterior marking, as specified in ASTM International's Standard D 4956[14]. (This is a medium-intensity "engineering grade" material using enclosed lens glass bead sheeting.).

8.4 INTERNATIONAL BUS REQUIREMENTS

ECE 36 and 107[38][39] require the following for bus emergency exit marking:

- An inscription reading "Emergency Exit" at every emergency exit, inside and out, and supplemented, where appropriate, by an international representative symbol.

- Markings on exit controls, inside and out, with a symbol or inscription.

- Clear instructions for the method of operation of exit controls.

ECE 36 also states that power-actuated doors must have illuminated signs or push buttons both inside and outside the coach to help passengers find the control in darkness.

ADR 44/02[41] requires conspicuous signs for all emergency exits with letters at least 25 mm (1 in) high and visible from the aisle. These signs must be either illuminated or "self-illuminating" for at least 15 minutes after loss of power from the vehicle battery. The door controls of service and emergency doors must be marked with "self-illuminating" material on the inside and retroreflective material on the outside. However, ADR 44/02 does not define specific performance criteria for either term.

ADR 58/00[42] does not include any mention of bus emergency exit signage.

British Regulation No. 257[43] requires that emergency exits be clearly marked as such, inside and outside the bus, and that the means of operation be clearly indicated.

8.5 RELEVANT RESEARCH

Although there is no known research that specifically addresses emergency exit signage and markings for motorcoaches or other buses, numerous studies and analyses have been conducted in other contexts. From this body of work, it is well understood that sign legibility depends on:

- Visual acuity of the observer (20/20, 20/40, etc.);
- Luminance of letters or symbols and their backgrounds;
- Distance from observer to sign;
- Angle between the observer's line-of-sight (LOS) and the plane of the sign;
- Obscuration by smoke or dust; and
- Dark adaptation.

Once these variables are specified quantitatively, there are established formulas to calculate required letter sizes. Extensive research has established the levels of illumination required to perform various tasks, such as reading characters of a specified size from a specified distance.[65] The 1983 NBS technical note report[44] was used by Volpe Center staff to prepare an illustration of letter size requirements as a function of distance, viewing angle, luminance, and luminance contrast. These calculations were used in establishing the minimum letter-size requirements for APTA industry standard passenger rail car emergency exit signs.

Viewing angle is important factor in sign legibility. To maintain legibility, when a sign is viewed off-axis, the letter size must increase as the aspect angle decreases (see Figure 8-11).

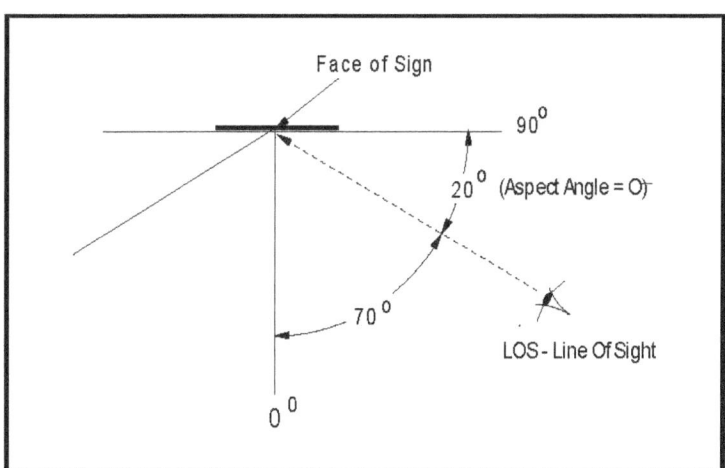

Figure 8-11. Illustration of Off-Axis Sign Viewing

To correct the calculation for off-axis viewing, it is necessary to divide the minimum size for on-axis viewing by the sine of the aspect angle. For a passenger more than one or two rows back from an emergency "exit" sign above the front door, the aspect angle can be very small, and the required letter size will be impractically large. A better approach to improve sign visibility is to use multiple emergency exit signs that include both longitudinal and transverse orientations.

Figure 8-12 shows that signs with 5 cm (2 in) letters can be read by persons with 20/40 vision up to a distance of about 2 m (7 ft), even when the luminance of the sign is only 0.3 foot-lamberts (~1 cd/m^2). (One cd/m^2 is typical of the initial luminance of HPPL signage.)

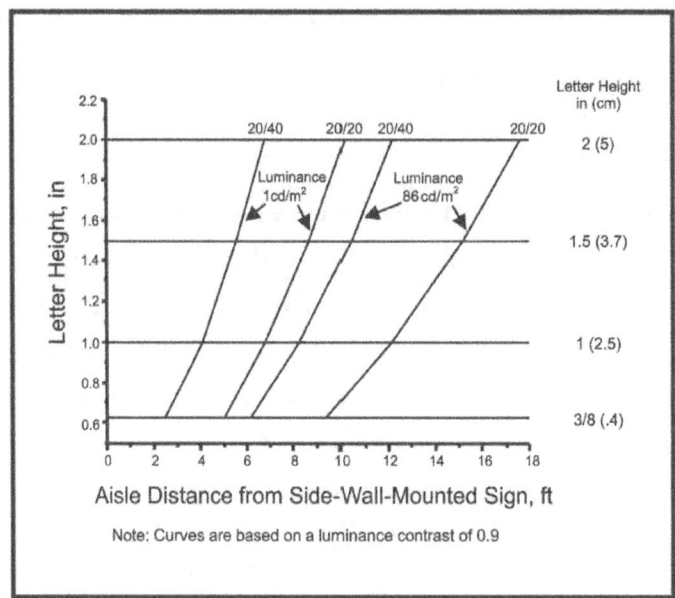

Figure 8-12. Letter-Size Requirements as a Function of Distance and Luminance

8.6 DISCUSSION

As noted earlier, NTSB has recommended that all motorcoach emergency exits be marked with signs using photoluminescent materials on the interior and retroreflective materials on the exterior ((H-00-01).[10]

Existing interior emergency exit signage currently installed on motorcoaches is legible in daylight or at night while the fluorescent lighting is in use. However, the conspicuity of some of the signage, such as the white-letters-on-black-background sign shown in Figure 8.3, is less than the other signs. At night, where reading lights are switched on, the signage on the emergency exit window release mechanisms is legible, but any decals located on the upper portion of the windows are not. In other nighttime lighting conditions, the conspicuity and legibility of the emergency exit signs required by FMVSS 217 ranges from marginal to nil. In buildings and other transportation modes, the issue of ensuring that emergency exit signs are clearly visible to individuals is usually addressed by the use of both a conspicuous sign located on, at or near the exit, with another such sign or marking located on, at, or near the emergency exit release handle, along with the opening instructions.

In addition, as noted in Chapter 2, FMCSA has issued "model" safety information brochures for distribution by bus operators to passengers that contain information including instructions and diagrams for opening the two types of motorcoach emergency exit windows (lift and pull) and emergency exit roof hatches.[27][28]

The use of photoluminescent (or other internally illuminated) material, or electrically-powered emergency lighting (with independent, self-contained power) to illuminate emergency exit signs and (and escape path marking) is required for most passenger aircraft, passenger rail cars, and ships. (Chapter 9 discusses emergency lighting more extensively.)

However, even with functioning, conspicuous emergency exit signage, motorcoach passengers may still encounter difficulty seeing and reaching emergency exits during nighttime or other dark conditions, if either the:

- Emergency signs and instruction on/above the exits are obscured by smoke, or
- Aisle is blocked or littered with crash or other debris.

The use of "low-location" emergency exit path marking systems, similar to those found on passenger aircraft and passenger rail cars, and on some European motorcoaches, could assist passengers to egress under these two conditions. However, because motorcoach exit path routes are much shorter than in other modes, less need may exist for installing such exit-path marking.

The FMVSS 217 school bus retroreflective marking performance criteria require a higher grade of material (Type III) than the FRA performance criteria (Type I, the lowest amount of reflectivity of the several ASTM grades). The FRA criteria recognizes that many passenger railroads use material that includes printed instructions with small letters on the surface of the marking, which would be hard to read with a flashlight if the higher grade of retroreflective material is used. FRA does not require outlining of the exterior door and window rescue access locations for emergency responders but does requires that those locations be identified by a conspicuous, unique, easily identifiable symbol. FRA requires that rail car roof access "soft spots," (and rescue access windows) be outlined on the exterior. Both NHTSA and FAA require that the exterior of all school bus and aircraft emergency exits, respectively, be outlined. However, since emergency responders will likely be unable to open motorcoach emergency exit windows from the outside due to FMVSS 217 window retention requirements, the need to mark the location of those exit locations on the motorcoach exterior is less essential.

8.7 CONSIDERATIONS

Potential motorcoach design changes that may increase the passenger egress rate and reduce the risk of injuries, as well as assist emergency responders to gain rescue access to the motorcoach interior during emergencies, include improved emergency exit signage and instructions consisting of:

- Interior
    - Increasing conspicuity
        - Place signage and instructions in location on or near the top or side of the exit that is more visible and of a color that contrasts with its background,
        - Use signs and instructions with larger minimum specific letter heights,
        - Use:
            * "High-performance photoluminescent" material,
            * Illumination by emergency lighting powered by crash-survivable, self-contained independent power sources, or
            * Dual-mode systems, that combine both technologies, and
    - Providing clearer, more easily understood instructions for passengers to release and open:
        - Front door, for emergency egress, if the driver is incapacitated, and
        - Emergency roof exit hatches, when the bus is overturned, and
- Exterior
    - Providing retroreflective signage and markings to identify location of emergency door exits, and emergency roof exit hatches, and
    - Providing instructions for opening emergency door exits and roof exits.

Additional guidance information relating to photoluminescent material used for motorcoach emergency exit signage is being developed by Volpe Center staff during the second year of this study.

9. EMERGENCY EXIT LIGHTING

According to the 2007 NHTSA research plan,[3] 28 out of 76 fatal bus crashes during 2001-2005 occurred (other than school buses) with the light condition listed as dark. During that same time period, 23 fatal motorcoach crashes took place, with 11 occurring in darkness and 2 occurring after dark but with other artificially lighted (e.g., streetlight) conditions. The NHTSA research plan stated that "Darkness has been found to be a relevant field condition in bus and motorcoach crashes" and included "illumination" as a priority research topic. NHTSA identified two potential strategies for improving nighttime illumination of emergency exit signage as: 1) photo luminescent signage, and 2) a backup electrical power source for illumination that is independent of the bus's electrical power system.

The 1997 Burnt Cabins, Pennsylvania motorcoach crash led NTSB to conclude that the lack of illumination was a major safety issue in bus crashes during darkness.[10] According to passengers, the motorcoach was "pitch black" after the crash, and they had difficulty finding the emergency exits. When emergency responders arrived, injured passengers were trapped within the bus and the interior was completely dark. The NTSB recommended that FMVSS 217 be revised to require that all motorcoaches be equipped with emergency lighting fixtures with self-contained independent power sources (H-00-01).

Some form of emergency lighting is required by regulation in the air, rail, and marine modes of common carriage of passengers. Emergency lighting is distinguished from normal lighting in that the lights are powered from independent batteries or other energy-storage devices, rather than engine-driven alternators. There has been a trend in other transportation modes toward requiring the emergency lights to have energy-storage devices that are independent of the main vehicle batteries, and usually contained within or immediately adjacent to the luminaire, so that crash damage cannot sever or short the power supply. (Note: "luminaire" is the term used in the lighting industry to describe the combination of a light fixture with its lamp(s), power supply, controller, and housing.)

9.1 FMVSS 217 REQUIREMENTS

The sole reference to emergency exit illumination in FMVSS 217 is the following:

> *S5.5.2 In buses other than school buses.* Except as provided in S5.5.2.1, each (exit) marking shall be legible, when the only source of light is the normal nighttime illumination of the bus interior, to occupants having corrected visual acuity of 20/40 (Snellen ratio) seated in the adjacent seat, seated in the seat directly adjoining the adjacent seat, and standing in the aisle location that is closest to that adjacent seat.

Although there seems to be an implied requirement for sufficient illumination that emergency exit signage will be legible, the lack of definitions of what is "normal" and precisely how legibility is to be measured have the effect of making this requirement unenforceable.

FMCSA regulations[66] require motorcoaches in fixed-route, interstate operations that transport persons with disabilities, to comply with ADA provisions for lighting, as contained in:[67]

> "(a) Any step well or doorway immediately adjacent to the driver shall have, when the door is open, at least 2 foot-candles of illumination measured on the step tread.
>
> (b) The vehicle doorway shall have outside light(s) which, when the door is open, provide at least 1 foot-candle of illumination on the pathway to the door for a distance of 3 feet (915 mm) to the bottom step tread or lift outer edge. Such light(s) shall be shielded to protect the eyes of entering and exiting passengers."

9.2 DESIGNS IN USE

On March 17, 2008, Volpe Center staff measured interior lighting levels in motorcoaches located at the Peter Pan Bus Lines maintenance facility in Chelsea, MA. Half of the tests were conducted on an *MCI* J4500 bus built in 2005, and the remainder on a *Van Hool* "2000" bus built in 1999. Tests were conducted after dark with the coaches parked in an open lot. Stray light from streetlights was negligible. These motorcoaches, along with those from other manufacturers, have three types of interior lighting:

- Bright, fluorescent lighting used when boarding and deboarding passengers,
- Individual reading lights above each seat, and
- Nightlights designed to provide only enough light for a passenger to walk down the aisle to the lavatory, so as to avoid interference with sleep and to avoid creation of reflections on the windshield.

The two motorcoach models tested have two rows of fluorescent lamps arrayed on the ceiling that are used during loading and unloading (see Figure 9-1) to provide a relatively high level of illumination. Illuminance measured on the floor at the center of the aisle ranged from 140 to 320 lux (13 to 30 foot-candles) and from 53 to 90 lux (5 to 8.4 foot-candles) at the release mechanisms of the windows.

These levels of illumination provided by the overhead fluorescents and the reading lights where used are sufficient to make all of the emergency exit signage readable and to charge photoluminescent emergency exit signage, installed at armrest level. However, the reading lights direct their output downward, so that they do not provide sufficient charging light for signs

Figure 9-1.   Typical Motorcoach – Bright Fluorescent Lighting

mounted at or near eye level on the window. Furthermore, these lights are nearly always off while the bus is in motion because they create reflections from the windshield and interfere with passenger desire for sleep.

The reading lights located above every seat produced an illuminance of 75 to 100 lux (7 to 9 foot-candles) directly below the lamp (see Figure 9-2). Stray light from these lights yielded readings of a few lux (~0.5 foot-candle) in the aisle or on the emergency exit window releases near seats where the reading lamp was switched on. Under these conditions, the signage on the window sill or midway up the emergency exit window was readable, but the "Emergency Exit" sign located on the emergency window exit surface near the top was nearly invisible. (However, FMVSS 217 does not require that additional emergency sign.) Typically, many reading lights are in use by motorcoach passengers in the early part of the night, but all or nearly all are switched off after midnight. Figure 9-2 shows the presence of a few small ceiling lights above the aisle, which are also on this circuit.

Figure 9-2.   Typical Motorcoach – Reading Lights On

Nightlights are the only lighting system that is always on at night. Their purpose is to provide just enough light for passengers to maintain spatial orientation or walk to the lavatory. The nightlight system is comprised of small incandescent or LED lamps in the ceiling and (in some models) under some seats. The floor-level illumination values measured by Volpe Center staff ranged from 0.1 to 0.8 lux (0.01 to 0.085 foot-candles). Illuminance at the emergency exit window release measured as much as 2 lux on one of the motorcoaches, but was only a small fraction of one lux on the other. At these levels, existing emergency exit signage is not legible. It is not possible to photograph these floor level lights without taking a long duration exposure, which would exaggerate their apparent brightness.

Some buses are equipped with "edge" lighting located on their steps (see Figure 9-3).

Figure 9-3.   Outline Lighting on Bus Step

9.3   REQUIREMENTS FROM OTHER MODES

Volpe Center staff reviewed emergency lighting regulations for other transportation vehicles and evaluated the adequacy of installed lighting systems that conform to those regulations from a human-factors perspective. All the other transportation modal regulations specify requirements for emergency lighting for emergency exit signs, except for passenger rail cars. For many years, NTSB recommended that passenger rail cars should be equipped with independently-powered door emergency door exit signage.

Principal requirements include:

- *Aviation* – 14 CFR, Subsection 25.812[29] – FAA requires illuminated emergency exit signs and specified levels of emergency lighting illuminance in various areas of the aircraft for aircraft seating 10 persons or more excluding pilot seats.

Red letters are required to be at least 3.8 cm (1.5 in) high on an illuminated white background that must have an area of at least 21 square inches (135 cm^2), excluding the letters. The letter height to stroke-width ratio may not be more than 7:1 or less than 6:1. The lighted background-to-letter contrast of the emergency exit signs must be at least 10:1. The emergency exit signs must be internally electrically illuminated with a background brightness (luminance) of at least 86 candelas per meter squared (cd/m^2) (25 foot-lamberts) and a high-to-low background contrast no greater than 3:1. Figure 9-4 shows an example of an aircraft emergency exit sign, powered by emergency lighting.

Figure 9-4. Aircraft Electrically-Illuminated Emergency Exit Sign

Passenger cabin general illumination must be provided so that the average illumination is not less than 0.05 foot-candle and the illumination at each 102-cm (40-in) interval is not less than 0.01 foot-candles, when measured along the centerline of main passenger aisle(s), and cross aisle(s) between main aisles, at seat arm-rest height.

A floor proximity exit path system is required to provide guidance for the emergency egress escape route, even when the overhead lights are obscured (as by smoke) (see Figure 9-5 and Figure 9-6).

FAA also requires that each door-operating handle marking must be conspicuous and be illuminated by the emergency lighting even under conditions of crowding.

The emergency lights and path marking must be powered by independent batteries and all components of the emergency lighting system must be able to withstand specified crash forces.

- *Rail* – 49 CFR, Subsection 238.115[30] – FRA requires that each new passenger rail car be equipped with an emergency lighting system provides an initial average illuminance of at least 1 foot-candle (10.76 lux), measured on the floor of the door location, which maintains operation for 90 minutes with no more than 40 percent reduction in illuminance, following loss of power from the main electrical system.

The back-up supply for the emergency lighting system must withstand specified crash forces.

Figure 9-5. Aircraft Floor Proximity Exit Path Marking – Electrically Powered
Note: Strip marking on left is not illuminated under non-emergency conditions)

 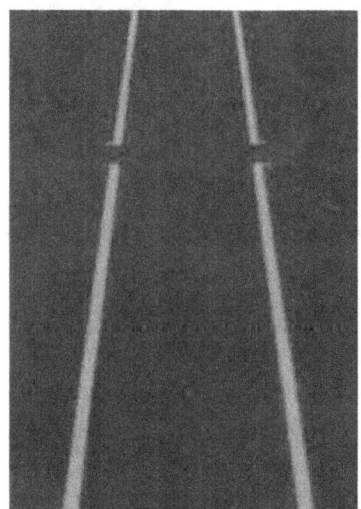

Figure 9-6. Aircraft Floor Proximity Exit Path Marking – Photoluminescent

In addition, Subsection 239.107[30] requires that passenger rail car exit doors used for emergency egress be identified either by (photo) "luminescent" signage (see Chapter 8), or (emergency) lighting located near the exits.

- *Marine* – SOLAS/IMO, Chapter 11-142[37] – Requires emergency lighting systems with self-contained power sources and 36-hour operation time for locations such as stairways, and exits.

The APTA industry standard for passenger rail car emergency lighting[32] requires that cars placed in service after 2012 have at least half of their emergency lights in each end of the car powered from independent energy-storage devices.

While the APTA passenger rail car emergency signage standard permits either electrically internally illuminated "high performance photoluminescent" emergency exit signs, or signs that are dual mode (which use both),[33] nearly all passenger railroad operators have chosen to install emergency exit signs and emergency exit locator signs comprised of "high performance photoluminescent" material, due to their lower maintenance and repair costs.

As noted in Chapter 8, FAA Advisory Circulars (AC) provide guidance to aircraft manufacturers and operators in evaluating systems to ensure that they comply with the floor proximity exit criteria, which requires an independent power supply or photoluminescent material. The first AC provides guidance to demonstrate that the system 1): enables each passenger to visually identify the escape path along the cabin aisle after leaving their seat, and 2) that the marking will enable the passenger to readily identify each exit from the exit path by referring only to marking and visual features when all illumination located 122 cm (4 ft above the floor) obscured by smoke, and it is dark.[63] The other AC provides guidance to demonstrate that photoluminescent material will provide the functional equivalent of electrically powered floor path marking.[64] The wording of the AC text indicates that "High Performance photoluminescent" material must be used; this material is permitted only for the path marking component, not exit signs.

The APTA passenger rail car "low-location" exit path marking standard permits either electrically-powered or "high performance photoluminescent" emergency exit signs and path marking, or dual mode signs that combine both technologies.[34] As noted above, passenger railroads have preferred to install "high performance photoluminescent" material, for these systems, again due to their lower maintenance and repair costs

As noted in Chapter 8, FRA is preparing a rulemaking[31] to incorporate the three APTA standards for emergency signage, low-location exit path marking, and emergency lighting, by reference into its regulations in the near future.

Lastly, FAA requires that exterior emergency lighting be provided at certain passenger aircraft emergency exit locations. At each over-wing emergency exit, the illumination must be not less than 0.03 foot-candles (measured normal to the direction of the incident light) on a 2-square-foot ($0.2 m^2$) area where passengers are likely to make their first step outside the cabin.

9.4 INTERNATIONAL BUS REQUIREMENTS

While ECE 36[38] requires artificial interior lighting for all aisles, steps and exit controls, it does not require "Emergency lighting" as such. However, it does require at least two interior lighting circuits with a stipulation that a failure in one cannot cause failure in the other, from which one may infer a requirement for redundant batteries. Furthermore, there is a requirement that the "emergency switch" (used to cut off the main battery) must not disable all interior lighting.

Neither ADR 44/02[41] nor ADR 58/00[42] contain any mention of emergency lighting.

British No. 257 regulation [43] does not explicitly address emergency lighting, but does require electric lighting to illuminate the path from every seat to the exits and especially on stairs. Redundant lighting circuits are required such that an electrical failure in one will not disable the other circuit.

9.5 RELEVANT RESEARCH

The University of Oklahoma Research Institute [19,23,24] conducted the only known research on the effect of illumination levels on motorcoach emergency egress in the 1970s. All of the trials in those three studies were conducted under bright light, so as to make motion-picture photography possible, but subjects wore dark goggles in some trials to simulate darkness or emergency lighting. Egress rates were usually somewhat higher during the trials with normal or emergency lighting (as opposed to darkness) but the differences were not statistically significant, because of the relatively small number of subjects in the trials.

It is unlikely that any controlled human factors experiments to measure subject egress rates from a motorcoach or a mock-up under conditions of total darkness would ever be approved by an IRB, because the risk of injury to subjects is too great.

However, during the development of FRA passenger rail car and APTA industry standards for emergency lighting, Volpe Center staff staged various demonstrations during which all passenger rail car lighting was turned off with participants seated in total darkness inside the car. These demonstrations were conducted using FRA and APTA participants at Penn Station in Philadelphia, PA; at Amtrak's Washington, DC maintenance yard; and at the "Rollover Rig,"[68] located at the WMATA rail transit car maintenance facility in Landover, MD. During the demonstrations, many participants became disoriented and several individuals stated that, in the absence of functioning emergency lights and/or a low-location exit path marking system to show the route to door emergency exits, the time required to evacuate a passenger rail car would significantly increase.

Another noteworthy finding of previous Volpe Center research, conducted for FRA, is that solid-state lighting systems (e.g., light-emitting diodes (LEDs)) offer numerous advantages over incandescent and fluorescent lamps for both normal and emergency lighting. These systems are greatly superior in terms of longevity, crash survivability, maintenance cost, compact size, energy efficiency, etc, and can charge photoluminescent signage more effectively than incandescent or normal fluorescent lamps. Despite their current initial high cost, these lighting systems are expected to replace other forms of lighting in most applications within the next decade.

9.6 DISCUSSION

Improving nighttime illumination of motorcoach emergency exit signage could be achieved by the use of photoluminescent signage; a backup electrical power source to provide illumination independent of the bus electrical power system; or a dual mode system combining both technologies.

As noted previously, NTSB has recommended that all motorcoaches be equipped with emergency lighting fixtures with self-contained independent power sources (H-00-01).[10]

Emergency lighting is necessary for passengers to maintain situational awareness during severe motorcoach crashes or other emergencies that occur at night or during other darkness conditions and to locate and use emergency exits when necessary.

Volpe Center staff field observations and measurements show that motorcoaches typical of the current U.S. fleet provide the following types of lighting:

- An overhead ceiling lighting system, used during boarding / deboarding, that provides an illuminance that complies with, and in fact, far exceeds the FMVSS 217 emergency-exit-sign legibility requirements, but that can be switched on only from the driver's console;

- An array of overhead reading lights, at each seat, that can render the emergency exit signage on adjacent window sills conspicuous and legible, if the passenger has switched the light on; and

- Nightlights that make it possible to see obstacles in the aisle, but do not provide sufficient illumination to render any of the emergency exit signage legible.

It is unclear which of these current lighting systems should be considered to be the "normal nighttime illumination" described in FMVSS 217. Certainly in any emergency situation that occurs at night or during other darkness condition in which it is possible to do so, the bus driver will switch on the bright ceiling lights. However, in a severe frontal crash or rollover, there are substantial risks that all of the lighting systems will become inoperable and that the driver may be incapacitated.

The illumination levels specified in the ADA requirements for door steps are double the requirements for passenger train emergency lighting and more than sufficient to make

"Emergency Exit" signs on the motorcoach doors conspicuous to passengers if the lights are mounted in the overhead ceiling. However, if the lights are installed in the step well and designed to illuminate only the treads and adjacent ground, the "Exit" signs located on or above the door may not be conspicuous to passengers at night or under other conditions of darkness. These lights are not required to function as part of a motorcoach emergency lighting system.

Other transportation modes have existing or proposed regulations that require the installation of emergency lighting systems with power sources independent of the vehicle's main battery, so that they have a reasonable probability of functioning after a crash. Solid-state-lighting systems (LEDs, etc.) are the preferred technology for achieving this objective, but their use is not widespread and has not yet been mandated.

The principal design shortcoming is not that existing bus lighting systems cannot provide sufficient illuminance, but rather that they cannot survive crash forces and, therefore, do not produce any emergency illumination at all following a motorcoach crash. Glass lamps in these systems are vulnerable to breakage and being dislodged from their sockets. In addition, the wiring supplying power to the lamps is vulnerable to being severed and/or shorted in the course of the structural deformations that occur in frontal collisions and rollovers.

Advances in lighting technology and energy storage devices have made it possible to build reliable, crash-survivable emergency lighting systems, which can be used to provide general emergency illumination or used as an integral component of lighted emergency exit signs.

Emergency lights with independent, self-contained energy-storage have been used successfully in buildings for decades and have been adapted for use in passenger rail cars. However, transportation vehicles differ from buildings in that their electric power source (the engine-driven alternator) is frequently shut off; this results in frequent full-discharge cycles of the backup battery in the emergency lighting and early battery failure. This issue can be avoided by either of the following approaches:

- A "smart" controller for battery-based emergency lighting that can prevent unnecessary activation of emergency lighting in normal engine-off conditions, while ensuring that it will be activated automatically whenever main power is lost unintentionally, or

- An independent power source for emergency lighting that is immune to failure from frequent discharge, such as a supercapacitor. (See example shown in Figure 9-7.) The unit provides an initial illuminance of at least 1 foot-candle in an oblong beam specifically shaped for aisle lighting. Supercapacitors in the power supply unit store sufficient energy to provide a rated run time of three hours after a ten-minute charge. The unit shown has averaged about 3.75 hours of operation on a full (10 minute) charge according to Volpe Center staff testing.

Figure 9-7.   LED Emergency Light with Capacitive Energy Storage

Human vision functions over an enormous range of lighting levels (one billion to one), and there is no general agreement regarding how much light an emergency system should provide. Various regulatory authorities specify emergency illuminance criteria ranging from 50 to 0.5 lux (5 to 0.05 foot-candles). Because an easily measured illuminance value is needed to determine compliance, yet there is no consensus number, some regulatory agencies, such as FRA, have selected a round number near the logarithmic center of the distribution, such as 10 lux or 1 foot-candle. FAA values are based on the available technology at the time the regulation was issued in 1967, given aircraft space and weight constraints.

Because buses are relatively small compared to passenger aircraft, trains, and ships, the U.S. requirements for their emergency illumination can be simpler, i.e., two emergency lights would be enough for passengers to locate and use the doors designated as emergency exits. An internally illuminated emergency exit sign with an integral "downlight," located at or over each service and other door, which could be used as an emergency exit, could provide both guidance to passengers for the location of the door emergency exit and situational awareness.

Volpe Center staff obtained and tested samples of the current state of lighting technology using commercially available LED emergency lights, suitable for installation in a motorcoach emergency exit sign, which produced illuminance values of 7 to 9 lux (0.6 to 0.8 foot-candle), as measured at the floor. The total light output can be as small as 20 lumens depending on the dispersion characteristics of the light diffuser. This light output results in a luminance value ranging from 1 to 100 cd/m^2, measured on the surface of the emergency exit sign through a photoluminescent overlay. This wide range of values is caused by varying dispersion characteristics of the LEDs and the transmission properties of the light cover or diffuser.

The minimum luminance needed to read an emergency exit sign from a distance of 3 m (10 feet), by a person with normal, dark-adapted vision is less than 10 mcd/m^2, i.e. less than one one-hundredth of the minimum luminance available on the signs measured, according to previous Volpe Center staff experiments conducted for FRA. Since it would be difficult to manufacture an internally illuminated emergency exit sign that generates an illuminance of 10 lux (1 fc) at the floor below it, without also producing a surface luminance of at least 1 cd/m^2, a potential minimum luminance criterion for such signs could be 1 cd/m^2.

9.7 CONSIDERATIONS

Potential motorcoach design changes that may increase the passenger egress rate and reduce the risk of injuries during emergencies include:

- Illuminated "Emergency Exit" signs with integral "downlights" located at or above each emergency exit, which provide a mean initial illuminance of at least 10 lux (1 fc) at the release handle or switch, whenever power for the normal lighting system is unavailable in an emergency.

- Each such designated emergency light with:
    - An independent energy-storage device (e.g., battery or supercapacitor) that can power it for at least one hour when power is unavailable from the main electrical system, as well as a circuit that will automatically activate the lamp in an emergency situation, i.e., upon unintentional loss of power from the main vehicle battery.
    - Energy storage capability sufficient to maintain the 10 lux (1 fc) illuminance for at least one hour after loss of power from the main electrical system.

During the second year of the Volpe Center study, other approaches to illumination of emergency exits and exit sign lighting, and exit path marking are being studied. Additional alternative technologies include exit sign identification using "high performance photoluminescent" (HPPL) material, or electrical power, as well as dual-mode systems, combining both.

10. FINDINGS, CONSIDERATIONS, AND FURTHER RESEARCH

10.1 SUMMARY OF FINDINGS

The literature review found that:

- Very little research focused on bus emergency egress has been conducted since Federally funded work was completed at the University of Oklahoma Research Institute in the 1970s.

- None of the existing research literature addresses egress through emergency exits currently installed in motorcoaches, which have sill heights and window weights much greater than those of buses tested in the 1970s.

- FMVSS 217 emergency exit requirements are different for school buses in various aspects than for other types of buses. Each school bus is required to have at least one emergency exit door and emergency exit identification requirements are more extensive.

- Other U.S. transportation regulatory agencies and industry standard specify requirements for emergency exits, including emergency exit identification and emergency lighting that could be adapted to apply to motorcoaches.

- The Economic Commission for Europe has established requirements for motorcoach emergency egress. These standards-include requirements for a second side-service or emergency door, larger emergency roof exit hatches than those required in the U.S., floor exit hatches, and marking of emergency exits and instructions for their operation on the bus exterior, as well as interior.

When necessary, safe and rapid passenger egress during a motorcoach emergency is possible when the front service door is available and none of the passengers have significant physical disabilities. When the front door is blocked or the motorcoach is overturned, egress is inherently slower and more challenging for passengers. However, if the front door is blocked, passenger emergency egress from a motorcoach may be more rapid if one or more of the following conditions exist:

- An additional usable side emergency exit door is available, located on either side of the motorcoach, such as a wheelchair-access door, or other additional side service door in the middle or the rear half of the motorcoach. These exits should be:
    - Conspicuously marked with emergency exit signs, as well as clear and understandable instructions for operation of the interior emergency release, and
    - Equipped with interlock and alarms systems to alert the driver if the exit is unlatched or open during operation.

- Usable emergency exit windows are available, the location for which is conspicuous to passengers. They should be designed so that typical passengers have the strength to open them, and provided with hold-open mechanisms and clear, understandable instructions for their operation.

- Usable emergency roof exit hatches are available, the location of which is conspicuous to passengers. Exit hatches should be designed so that they can be easily opened and provided with clear instructions illustrating their use in an overturned condition.

Volpe Center staff conducted a series of experimental trials to generate preliminary egress time estimates for a fully loaded 56-seat motorcoach in daylight, as shown in Table 10-1. These estimates were based on three assumptions:

- All passengers have sufficient knowledge and agility to traverse the various egress paths.

- The side door (a wheelchair-access door in these experiments) is unobstructed by seats and fitted with an internal exit release.

- For the purpose of these estimates, the emergency exit windows are assumed to be fitted with "hold open" mechanisms.

Table 10-1. Volpe Center Preliminary Motorcoach Egress Estimate – 56 Passengers

EGRESS PATH	NUMBER OF EXITS. USED	OPENING TIME (min)	FLOW RATE (exit/ppm)	EGRESS (min)	TOTAL (min)
Front door	1	.05	36	1.56	1.61
Windows	6	.2	9	1	1.20
Wheelchair-access door	1	.2	25	2.24	2.44
Roof hatch	2	.1	12	2.33	2.43

Motorcoach passenger egress times are likely to be considerably higher than those in Table 10-1 in actual emergencies, if one or more of the following conditions apply:

- The driver is incapacitated and cannot assist passengers;

- Some passengers are injured;

- Some passengers do not understand how to open or use the emergency exits;

- Some passengers lack the agility or strength to use the emergency exits;

- Passengers cannot find or see how to use the exits because of darkness and/or smoke; and/or
- Emergency exits will not stay open.

Volpe Center staff conducted field measurements of illumination and letter sizes for motorcoach emergency exit sings showed that those in current use comply with FMVSS 217 requirements, whenever there is at least a low level of illumination present (i.e., in daylight or when the fluorescent boarding lights or adjacent reading lights are in use). However, the typical level of illumination provided during the night, 0.1 to 0.8 lux (0.01 to 0.085 fc), by the "night lights") does not allow the exit signage to be conspicuous or easily legible, even at very short viewing distances.

10.2 SUMMARY OF PRELIMINARY DESIGN CHANGE CONSIDERATIONS

Issues relating to barriers to rapid motorcoach emergency egress have been documented in various NTSB reports and OKRI research reports. The results of the Volpe Center study conducted to date are consistent with the findings contained in those reports.

Other U.S. transportation regulatory agency requirements for vehicle emergency exits, including exit identification, and emergency lighting, could be adapted for application to motorcoaches. These requirements (extensively described in the Year 1 interim report) include: more than one emergency exit door, larger emergency roof exit hatches, photoluminescent marking of emergency exits on the interior, retroreflective marking of emergency exits on the exterior, and independently-powered emergency lighting.

Certain provisions of existing FMVSS 217 requirements for school bus emergency exits and standards established by the Economic Commission for Europe for motorcoaches operated in other countries could also be adapted and applied to motorcoaches.

European Union for motorcoaches operated in other countries could also be adapted and applied to motorcoaches.

Potential motorcoach design changes that may increase the passenger egress rate and reduce the risk of injuries during emergencies are:

- Positive "hold open" devices for:
  o Doors that can be used for emergency egress,
  o Emergency exit windows, and
  o Emergency roof exit hatches;

- Improved emergency exit signage and instructions:
  - Interior
    - Increase conspicuity:
      * Place signage in location on or near the top or side of the exit that is more visible and a color, that contrasts with its background,
      * Use signage and instructions with larger minimum specific letter heights, and
      * Use:
        ~ "High-performance photoluminescent" material,
        ~ Illumination by emergency lighting powered by crash-survivable, self-contained independent power sources, or
        ~ Dual-mode systems, that combine both technologies,
    - Provide clearer, more easily understood instructions for passengers to release and open:
      * Front door, for emergency egress, if the driver is incapacitated.
      * Emergency roof exit hatches, when the bus is overturned, and
  - Exterior
    - Provide retroreflective signage and markings to identify location of door emergency exits and emergency roof exit hatches, and
    - Provide instructions for opening door emergency exits and emergency roof exit hatches;
- Increased minimum number and size of emergency roof exits:
  - At least two hatches per motorcoach, and
  - Larger aperture dimension of 4,000 cm^2 (620 in^2);
- An additional floor-level door exit on the side, located in the middle / rear half of the bus for use by passengers in an emergency, by either:
  - Modification of the wheelchair-access door to permit it to be opened from inside for use as an emergency exit, and /or
  - Another door exit that could be opened for use as an emergency exit.

10.3 FURTHER RESEARCH

Volpe Center staff are investigating the following motorcoach emergency egress topic items in Year 2 of this study:

- Conduct of human factors experiments to evaluate:
  - Human strength of a population of adult subjects evenly balanced by age and gender to apply the pulling and pushing forces needed to open exit doors and top-hinged emergency exit windows,

- Effects of illuminance levels on adult egress rates,
- Rates of egress through:
  - Wheelchair-access doors with two different configurations and clearances, and
  - Stairways with 30 cm (12 in) step risers similar to those used for the second service door on many buses operated in other countries; and

• Development of appropriate performance specification criteria for:
  - Photoluminescent emergency exit marking and sign materials, including luminance, dimensions, and contrast requirements,
  - Electrically powered-illuminated emergency exit signs, and
  - Dual-mode exit marking systems, combining both technologies.

11. REFERENCES

1. National Highway Traffic Safety Administration (NHTSA). *Title 49, Transportation, Code of Federal Regulations (49 CFR). Part 571, Subpart B, Federal Motor Vehicle Safety Standards (FMVSS)*. U.S. Department of Transportation (USDOT). National Archives and Records Administration, Washington, DC. As of October 1, 2008.

2. NHTSA. *49 CFR,. Part 571, (FMVSS), § 571.217, Standard No. 217, Bus Emergency Exits and Window Retention and Release*. USDOT. National Archives and Records Administration, Washington, DC. As of October 1, 2008.

3. NHTSA. *NHTSA's Approach to Motorcoach Safety*. Docket 2007-28793. August 6, 2007. Office of Crashworthiness Standards. NHTSA/USDOT. Washington, DC.

4. National Transportation Safety Board (NTSB). *Most Wanted List: Enhanced Protection for Motorcoach Passengers*. November 2007.

5. NHTSA. *49 CFR, Part 571, Subsection 571.3.Definitions*. USDOT. National Archives and Records Administration, Washington, DC. As of October 1, 2008.

6. Federal Motor Carrier Safety Administration (FMCSA). *49 CFR,. Parts 300-399, Federal Motor Carrier Safety Standards (FMCSS)*. USDOT. National Archives and Records Administration, Washington, DC. As of October 1, 2008.

7. National Transportation Safety Board (NTSB). *Motorcoach Fire on Interstate 45 during Hurricane Rita Evacuation near Wilmer, TX, September 23, 2005*. Report no. NTSB-HAR 07/01. Adopted February 21, 2007.

8. NTSB. *Selective Motorcoach Issues*. Report no. NTSB/SIR–99/01. Adopted February 11, 1999.

9. NTSB. *Bus Crashworthiness Issues*. Report no. NTSB/SIR–99/04. Adopted September 21, 1999.

10. NTSB. *Greyhound Motorcoach Run-Off-the Road Accident, Burnt Cabins, Pennsylvania, June 20, 1998*. Report no. NTSB-HAR-00/01. Adopted January 5, 2000.

11. FMCSA. *49 CFR., Part 393 Parts and Accessories and Part 396 Inspection, Repair, and Maintenance*. USDOT. National Archives and Records Administration, Washington, DC. As of October 1, 2008.

12. NHTSA. "Final Rule. 49, CFR, Part 571 Federal Motor Vehicle Safety Standards, Bus Retention and Release. Subsection 571.217." *Federal Register*, Rules and Regulations. May 10, 1972, Vol. 37, No. 91. 9394-9397. Rules and Regulations. USDOT. National Archives and Records Administration, Washington, DC.

13. NHTSA. *49 CFR,. Part 571, FMVSS 220, School Bus Rollover Tests.* USDOT. National Archives and Records Administration, Washington, DC. As of October 1, 2008.

14. ASTM International. *ASTM D 4956-07e1, Standard Specification for Retroreflective Sheeting for Traffic Control.* West Conshohocken, PA. 2007.

15. NHTSA. *49 CFR. Part 571, (FMVSS), § 571.125, Standard No.125, Warning Devices.* USDOT. National Archives and Records Administration, Washington, DC. As of October 1, 2008.

16, FMCSA. *49 CFR. FMCSS, Part 393; Subsection 393.62, Emergency Exits for Buses.* USDOT. National Archives and Records Administration, Washington, DC. As of October 1, 2008.

17. FMCSA. *49 CFR. FMCSS, Part 396; Subsection 396.3, Inspection, Maintenance and Repair, (a) (2).* USDOT. National Archives and Records Administration, Washington, DC. As of October 1, 2008.

18. Interstate Commerce Commission (ICC). *Final Rule. Subchapter B – Carriers by Motor Vehicle; Parts 190-197, Safety Regulations. Subpart D, Glazing and Window Construction; Part 193, Subsection 193.61, Window Construction; 193.62, Window Obstructions; 193.63 Window markings; and Subpart G, Miscellaneous Parts and Accessories, 193.92, Buses, marking emergency doors. Federal Register,* May 15, 1952, Vol. 17. 4422-4453. Rules and Regulations. National Archives and Records Administration, Washington, DC.

19. Purswell, J.L.; Dorris, A.L.; and Stephens, R.L. *Escapeworthiness of Vehicles and Occupant Survival.* Prepared for FHA/NHTSA/USDOT by School of Industrial Engineering, University of Oklahoma Research Institute (OKRI). Norman, OK. Final Report. Part 1, Report no. DOT-HS-800-428. Parts 2 and 3, Report no. DOT-HS-800-429. December 1970.

20. MARTEC. *Window Impact Forces from Occupants During Motor Coach Rollover.* Prepared for Transport Canada, Safety and Security, Road Safety and Motor Vehicle Regulations. Technical Report TR04-41, Rev 4, June 2004. *Canada.*

21. NHTSA. *Regulatory Review Assessment; Federal Motor Vehicle Safety Standard FMVSS No. 217, Emergency Exits and Window Retention and Release.* June 2007.

22. Smith, R.; Tobe, J.; et al. *Engineering Assessment of Current and Future Vehicle Technologies: FMVSS No. 217 Bus Emergency Exits and Window Retention and Release and FMVSS No. 220, School bus Rollover Protection.* Prepared for NHTSA/USDOT by Battelle, Inc.; University of Michigan Transportation Research Institute (TRC), Inc.; and Smithers Scientific Services, Inc. Contract number DTNh22-02-D-02104, Task Order No. 04. December 2005.

23. Sliepcevich, C.M.; et al. *Escapeworthiness of Vehicles for Occupancy Survivals and Crashes*. Prepared for NHTSA/USDOT, by School of Industrial Engineering, OKRI. Norman, OK. Final Report. Part 1: Report no. DOT-HS 800 736. Part 2: DOT-HS 800-737. July 1972.

24. Purswell, J.L.; Dorris, A.L.; and Stephens, R.L. *Evacuation of Intercity Buses*. Prepared for Federal Highway Administration/USDOT, by School of Industrial Engineering, OKRI. Norman, OK. Final Report. January 1978.

25. Van Cott, H.P. and Kinkade, R.G. *Human Engineering Guide to Equipment Design*. U.S. Joint Army Navy Steering Committee. 1972.

26. FMCSA. "Pre-Trip Safety Information for Motorcoach Passengers." *Federal Register*, Notices. August 28, 2007, Vol. 71, No. 166. 50971-50973. USDOT. National Archives and Records Administration, Washington, DC.

27. FMCSA *Safety Information for Bus/Motorcoach Passengers, What you Should Know. Bus Lift*. Brochure – FMCSAES0-08-0006. August 2008. Website: http://www.fmcsa.dot.gov/documents/outreach/bus/PreTrip-SafetyBrochure-Lift.pdf.

28. FMCSA *Safety Information for Bus/Motorcoach Passengers, What you Should Know. Bus: Handle*. Brochure – FMCSAES0-08-0011. August 2008. Website: http://www.fmcsa.dot.gov/documents/outreach/bus/PreTrip-SafetyBrochure-Pull.pdf.

29. Federal Aviation Administration (FAA). *14 CFR. Part 25, Subparts 25.803, Emergency Evacuation; 25.807, Emergency Exits; 25.809, Emergency Exit Arrangement; 25.810, Emergency Egress Means and Escape Routes; 25.811 Emergency Exit Marking; 25.812, Emergency Lighting; and 25.813, Emergency Exit Access*. USDOT. National Archives and Records Administration. Washington, DC. As of January 1, 2009.

30. Federal Railroad Administration (FRA). *49 CFR. Part 238, Passenger Equipment Safety Standards; and Part 239, Passenger Train Emergency Preparedness*. FRA / Department of Transportation (FRA/USDOT). Office of the Federal Register, National Archives and Records Administration. As of October 1, 2008.

31. FRA. "Final Rule. 49, CFR, "Parts 223 and 238, Passenger Train Emergency Systems; Emergency Communication, Emergency Egress, and Rescue Access." *Federal Register*, Rules and Regulations. February 1, 2008, Vol. 73, No. 22. 6370-6413. USDOT. National Archives and Records Administration, Washington, DC.

32. American Public Transportation Association (APTA). *APTA SS-013-99, Rev. 1, Standard for Emergency Lighting System Design for Passenger Cars*. Approved September 11, 2007 and authorized October 7, 2007. Standard originally approved March 4, 1999 and authorized March 17, 1999. *Manual of Passenger Rail Equipment Safety Standards* (PRESS). Washington, DC.

33. APTA. *APTA SS-PS-002-98, Rev.3, Standard for Emergency Signage for Egress/Access of Passenger Rail Equipment.* Revision 3 approved September 11, 2007 and authorized October 7, 2007. Stadnard originally approved June 15, 1998. PRESS. Washington, DC.

34. APTA. *APTA SS PS-004-98, Rev. 2, Standard for Low-Location Exit Path Marking.* . Revision 2 approved September 11, 2007 and authorized October 7, 2007. Standard originally approved May 21, 1999 and authorized October 7, 1999. PRESS. Washington, DC.

35. United States Coast Guard (USCG). *46 CFR. Shipping, Subchapter H, Passenger Vessels and Subchapter T, Small Passenger Vessels.* Department of Homeland Security. National Archives and Records Administration. Washington, DC. As of October 1, 2008.

36. USCG. *Navigation and Vessel Inspection Circular No. 8-93, Equivalent Alternatives to 46 CFR, Subchapter H, Requirements Related to Means of Escape, Safe Refuge Areas, and Main Vertical Zone Length* (NVIC-8-93). Commandant, Office of Marine Safety Security and Environmental Protection, Ship Design Branch, Safety and Oversight Section (G-MTH-4). November 1993.

37. International Maritime Organization (IMO). *International Convention of Safety to Life at Sea (SOLAS).* Consolidated Edition, 2004. / IMO Resolution A.752(18) *Guidelines of Evaluation, Testing and Application of Low Location Lighting on Passenger Ships.*

38. Economic Commission for Europe (ECE). *ECE No. 36: Uniform Provisions Concerning the Approval of Large Passenger Vehicles with Regard to their General Construction. Addendum 35, Revision 3; February 20. 2008.* Including: Addendum35, Revision 3, Amendment 4, 14 December 2004; Addendum 35, Revision e, Amendment 3; Addendum35, Revision 3, Amendment 2, 18 December 2003; and other previous supplements.

39. ECE. *ECE No. 107: Uniform Provisions Concerning the Approval of Categories of M_2 and M_3 Vehicles with Regard to their General Construction.* Revision 2, entered into force October 1995. Addendum 106, Revision 1, 8 October 2004. Including: Addendum 106, Revision 1, Amendment 1, January 2008; Addendum 106, Revision 1, Erratum 1, January 2006; and Addendum 106, Revision 1, Corrigendum 1, October 2004, and other previous supplement.

40. Transport Canada. *Review of Bus Safety Issues.* TP 133330E. November 1998. *Canada.*

41. Australian Design Rules (ADR) *44/02, Requirements for Omnibuses Designed for Hire and Reward.* And previous versions 44/00 and 44/01. National Transport Commission. 1993. *Australia.*

42. *ADR 58/00, Requirements for Omnibuses Designed for Hire and Reward.* National Transport Commission. 2006. *Australia.*

43. Department of Transport. No. 257, *Public Service Vehicles (Conditions of Fitness, Equipment, Use and Certification) Regulation.* 1981. *United Kingdom.*

44. Howett, G.L. *Size of Letters Required for Visibility as a Function of Viewing Distance and Observer Visual Acuity.* Prepared by Building Physics Division, National Engineering Laboratory, National Bureau of Standards (now National Institute of Standards and Technology (NIST), Department of Commerce. Prepared for Occupation and Safety and Health Administration, U.S. Department of Labor. Technical Note 1180. July 1983.

45. *MIL STD 1472F Design Criteria Standard for Human Engineering.* August 29, 1999; Superceding MIL-STD-1472E. Department of Defense. Washington, DC.

46. Government Consumer Safety Research. *Strength Data for Design Safety. Phase 1.* Prepared for Department of Trade and Industry. 2002. *United Kingdom.*

47. Government Consumer Safety Research. *Strength Data for Design Safety. Phase 2.* Prepared for Department of Trade and Industry. 2002. *United Kingdom.*

48. TCRP Report 100. *Transit Capacity and Quality of Service.* Transit Cooperative Research Program. Transportation Research Board. 2nd edition. 2003.

49. Machek, E.C., et al. *Mendenhall Glacier Visitor Center Vehicular and Pedestrian Traffic Congestion Study.* Prepared by Volpe Center/USDOT for U.S. Forest Service, Department of Agriculture. Final Report. Report no. DOT-VNTSC-USDA-07-01. May 2007.

50. NHTSA. *Laboratory Test Procedure for FMVSS 217.* TP-217TB-00. June 2002.

51. FMCSA. *Basic Plan for Motorcoach Passenger Safety Awareness.* http://www.fmcsa.dot.gov/about/outreach/bus/bus-safety-awareness-plan.htm. 2008.

52. FRA. *Research Results: Passenger Rail Car Egress Time Prediction.* RR-06-04. May 2006. Washington, DC. Website: http://www.fra.dot.gov/downloads/Research/rr0604.pdf.

53. Fujiyama, T.C. and Tyler, N. *An Explicit Study on Walking Speeds of Pedestrians on Stairs.* Center for Transportation Studies, University College, London. January 2004. *United Kingdom.*

54. Fruin, J.J. *Pedestrian Planning and Design.* Metropolitan Association of Urban Designers and Environmental Planners. New York. 1971.

55. Markos, S.H. and Pollard, J.K. *Passenger Train Emergency Systems: Single Level Commuter Rail Car Egress Experiments.* Prepared by Volpe Center / USDOT for FRA/US DOT. (Currently in FRA approval process for publishing in 2009).

56. ECE, Inland Transport Committee World Forum for Harmonization of Vehicle Regulations (WP 29). *Group of Experts on General Safety (GRSG) Regulation 107 (M2 and M3 vehicles).* Proposal for amendments concerning emergency windows. Informal Document no. GRSG. 94-02. 94th Session of GRSG, 21-24. April 2008.

57. McLean, G.A. and C. Corbet. *Access to Egress III: Repeated Measurement of Factors That Control the Emergency Evacuation of Passengers Through the Transport Airplane Type-III Overwing Exit.* Prepared by the Civil Aerospace Medical Institute (CAMI), Office of Aviation Medicine, FAA/USDOT. Final report. Report no. DOT/FAA/AM-04/2. January 2004.

58. McLean, GA., et al. *Aircraft Evacuations Through Type-III Exits I: Effects of Seat Placement at the Exit.* Prepared by CAMI, Office of Aviation Medicine, FAA/USDOT Final report. Report no. DOT/FAA/AM-95/22. July 1995.

59. McLean, GA., et al. *Access to Egress I: Interactive Effects of Factors That Control the Emergency Evacuation of Naïve Passengers Through the Transport Airplane Type-III Overwing Exit.* Prepared by CAMI, Office of Aerospace Medicine, FAA/USDOT. Final Report. Report no. DOT/FAA/AM-02/16. August 2002.

60. McLean, G.A. Access to Egress: *A Meta-Analysis of the Factors that Control Emergency Evacuation through the Transport Airplane Type III Overwing Exit.* Prepared by CAMI, FAA/USDOT for Office of Aviation Medicine, FAA/USDOT. Final report. Report no. DOT/FAA/AM-01/2. January 2001.

61. ECE, Inland Transport Committee World Forum for Harmonization of Vehicle Regulations (WP 29). *Group of Experts on General Safety (GRSG) Regulation 107 (M2 and M3 vehicles).* Proposal for amendments concerning emergency windows. Informal Document no. GRSG. 94-02. 94th Session of GRSG, 21-24. April 2008.

62. NHTSA. *Motorcoach Roof Crush Rollover Testing.* (Tests conducted at MGA, Inc., Burlington, WI. February 2008). Discussion paper. Washington, DC. March 2009.

63. FAA. Advisory Circular (AC) 25.812 1A. *Floor Proximity Emergency Escape Path Marking.* Air Certification, FAA/US DOT. May 22, 1989. Washington, DC

64. FAA. AC 25.812 2. *Floor Proximity Emergency Escape Path Marking Incorporating Photoluminescent Elements.* Air Certification, FAA/USDOT. July 24, 1997. Washington, DC.

65. Moon, B. *The Scientific Basis of Illuminating Engineering.* Revised Edition. Dover Publications, New York. 1961.

66. FMCSA. *49 CFR. Part 374, Passenger Carrier Regulations, Subpart C, Adequacy of Intercity Motor Common Carrier Passenger Service, Subsection 374.315, Transportation of Passengers with Disabilities.* Office of the Secretary. USDOT. National Archives and Records Administration, Washington, DC. As of October 1, 2009.

67. *49 CFR, Part 38. Americans with Disabilities Act (ADA) Accessibility Specifications for Transportation Vehicles, Subsection 38.31, Lighting.* Office of the Secretary. USDOT. National Archives and Records Administration, Washington, DC. As of October 1, 2009.

68. FRA. *Research Results: Passenger Rail Car Evacuation Simulator.* RR-06-04. April 2006. Washington, DC. Website: http://www.fra.dot.gov/downloads/Research/rr0607.pdf.

APPENDIX A. U.S. Bus Emergency Egress Regulation History

AGENCY	FMVSS / FMCSS	FEDERAL REGISTER DATE	TYPE OF NOTICE (PROPOSED OR FINAL RULE)	REQUIREMENTS / SUBJECT	EFFECTIVE DATE
ICC	194	1937	12-23-1936	Necessary Part and Accessories – did not include any mention of bus emergency exits; however, other emergency equipment, including first aid kit for bus with more than 10 persons was required	July 1, 1937
ICC	193	12/14/1946	Proposed Rule	Initial general notices	NA
ICC	193	1/3/1951	Proposed Rule	Revised definition of bus to include taxi-cabs. Windows: Three elements: window construction, including unobstructed area for means of escape and push-out windows; lack of barriers that interfere with the unobstructed area size; and window markings. Emergency Exit Doors: Marking with 1 in letter and lighted by a red light	NA
ICC	193	5/15/1952	Final Rule	Adopted proposed rule text.	May 15, 1952
Bureau of Highway Safety (BHS) / Dept of Commerce (DOC)	255	12/3/1966	Proposed Rule	Initial FMVSS regulations. Although the subject of bus window retention and emergency exits was not originally included in the FMVSS regulations, definitions for "buses" and "school buses" were included.	NA
BHS / DOC	255	2/3/1967	Final Rule	Initial FMVSS regulations were adopted. Bus window retention and emergency exits not included.	January 1, 1968.
BMSC	293	12/15/1967	Final Rule	Renumbering to reflect move from ICC	December 15, 1967

A-1

U.S. Bus Emergency Egress Regulation History (2)

AGENCY	FMVSS / FMCSS	FEDERAL REGISTER DATE.	TYPE OF NOTICE (PROPOSED OR FINAL RULE)	REQUIREMENTS / SUBJECT	EFFECTIVE DATE
FHWA / BHS	255	10/14/1967	Proposed Rule	Add new rule to secure installation of bus side and rear windows, and the prohibition of push-out windows. In addition, the inclusion of requirement for readily accessible emergency windows operated from both inside and outside the bus, capable of being actuated with a minimum of effort consistent with containment effectiveness, was proposed. Comments were requested related to emergency exit criteria, such as operating mechanisms, method of mounting, size and location	FHWA / BHS
FHWA / Bureau of Motor Carrier Safety (BMCS)	393	12/25/1968	Final Rule	New 371 Moved from 255 293 renumbered 393	12/25/1968
NHSB / FHWY		8/15/1970	Proposed Rule	Proposed side and rear windows retention requirements and that buses other than school buses be required to be equipped with push-out windows providing unobstructed openings, the size for which related to the number of seating positions. Push-out windows would also be required to meet accessibility, force limits, opening dimensions, and marking to assure identification and operation as emergency exits.	NA
BMCS	393	8/15/1970	Proposed Rule	Revise provisions to make consistent with substance of anticipated new FMVSS 217 requirements	NA

A-2

U.S. Bus Emergency Egress Regulation History (3)

AGENCY	FMVSS / FMCSS	FEDERAL REGISTER DATE.	TYPE OF NOTICE (PROPOSED OR FINAL RULE)	REQUIREMENTS / SUBJECT	EFFECTIVE DATE
National Highway Safety Bureau (NHSB / FHWA	217	10/14/1970	Proposed Rule	Purpose: To provide requirements for bus window retention and "push-out" windows" used as emergency exits. Would have required the secure installation of side and rear windows, including large buses such as motorcoaches, and transit buses. In addition, in contrast to earlier 1967 proposal, proposed rule would have required push out emergency windows with accessibility, force limits, opening dimensions, and marking to assure adequate identification and adequate instructions for emergency exit operation. (Requirements for mandating push-out windows for school buses were not proposed.)	NA
NHTSA	217	5/10/1972	Final Rule	"Bus Window Retention and Release" Certain clarifications were made to the 1970 proposed rule, including: the added option to use other than push-out windows, such as doors and panels, meeting emergency exit requirements; permitting of an alternative roof hatch instead of a rear door to provide design flexibility while providing for emergency egress in rollover situations; higher force requirements for exit release to improve latch integrity; an exemption from the exit opening sizes for smaller buses, and clearer labeling provisions. Push-out windows or other emergency exits were not required for school buses due to the risk of children falling from the windows; however if such windows or were installed, they were required to meet emergency exit requirements.	September 1, 1973

U.S. Bus Emergency Egress Regulation History (4)

AGENCY	FMVSS / FMCSS	FEDERAL REGISTER DATE	TYPE OF NOTICE (PROPOSED OR FINAL RULE)	REQUIREMENTS / SUBJECT	EFFECTIVE DATE
FHWY / BMCS	393	6/10/1972	Final Rule	Adopted 1970 proposed revisions including some clarifications. Buses built before September 1, 1973 were permitted to comply with the earlier 1968 regulations or FMVSS 217, at the option of the operator.	July 1, 1973
NHTSA	217	9/6/1972	Final Rule	Amendments including permission to provide at least one door per three persons to meet the unobstructed opening requirement; as well as clarifying the window retention provision to allow the use of tempered glass for emergency exit windows.	September 1, 1973
NHTSA	217	3/6/1973	Final Rule	Response to petitions for reconsideration and amendments included deletion of torque requirement for emergency windows, as well as revised and added certain figures to make clear that exit access areas, including roof exits, must be such that the occupant has access when the bus is upright or on its side.	September 1, 1973
NHTSA	217	3/23/73	Correction	A correction was made in March 1973 to correct a transposition of figure diagrams.	March 23, 1973
NHTSA	217	10/1/1973 5/2/1974	Proposed Rule Final Rule	At request of DOJ, proposed to exempt buses that transport persons under physical restraint (e.g., prisoners). In May 1974, NHTSA issued a notice adding that exemption.	June 4, 1974

U.S. Bus Emergency Egress Regulation History (5)

AGENCY	FMVSS / FMCSS	FEDERAL REGISTER DATE.	TYPE OF NOTICE (PROPOSED OR FINAL RULE)	REQUIREMENTS / SUBJECT	EFFECTIVE DATE
NHTSA	217	2/28/1975	Proposed7 Rule	Proposed to make it mandatory for school buses to contain emergency exits and to meet additional requirements, such as installing at least one rear emergency door or at least 2 side emergency doors; and that door release mechanisms operate without the use of remote controls or tools, be connected to the engine ignition to prevent operation, and to sound a warning when they are open and the release is unlatched. In addition, the marking of the emergency doors was proposed to require to state "Emergency Doors," using 2-inch high letters with a contrasting color; as well as arrows to indicate which direction the release would be located, adjacent to the exit, again in a contrasting color	NA
NHTSA	220	2/28/1975	Proposed Rule	Proposed a new rule for school bus rollover protection. The intent was to provide structural integrity to school bus passenger compartment during rollover forces. The proposed rule included a requirement to require emergency exit operability during and after the force application to the roof.	NA
NHTSA	217	4/18/1975	Proposed Rule	Proposed to amend requirements to permit certain types of non-push-out manually operated emergency exits to be unmarked, in buses having a GVWR of less than 10,000 pounds.	NA
NHTSA	217	5/28/1975	NA	Extended the comment period for the proposed school bus requirement revisions from April 29, 1975	June 26, 1975

A-5

U.S. Bus Emergency Egress Regulation History (6)

AGENCY	FMVSS / FMCSS	FEDERAL REGISTER DATE.	TYPE OF NOTICE (PROPOSED OR FINAL RULE)	REQUIREMENTS / SUBJECT	EFFECTIVE DATE
NHTSA	217	10/16/1975	Final Rule	Decision that special markings for emergency exits are not necessary for doors and manually operated windows in small buses; however, the size requirements for unobstructed exit openings must be complied. Specially installed emergency windows, such as push-out windows must still be marked as designated emergency window	October 16, 1975
NHTSA	571.3	12/31/1975	Final Rule	Amends the definition of school bus to conform with 1974 Act by expanding present definition.	October 27, 1976
NHTSA	217	1/27/1976	Final Rule	Amended to include specific requirements for school bus emergency doors with minimal revisions from the April 1975 notice	October 27, 1976
NHTSA	220	1/27/1976	Final Rule	New rule required school bus emergency exits to be operable, i.e., opening as specified in FMVSS 217, during and after the roof force application. The original proposed requirement to close the emergency exits after the force was removed was deleted.	October 27, 1976
NHTSA	217	1/27/1976	Proposed Rule	Proposed to amend FMVSS 217 to permit the use of a rear window emergency exit and a side door emergency exit, (in combination) as an optional means of compliance with the school bus emergency exit requirements, or to permit that option only in rear engine buses. Proposed clarifications to the emergency exit labeling and instructions for all bus emergency exits were intended to provide guidance regarding the location of emergency exits and the actions necessary to release and open the exits. In addition, specific letter height, color and contrast provisions for designated school bus emergency exits were proposed.	NA

U.S. Bus Emergency Egress Regulation History (7)

AGENCY	FMVSS / FMCSS	FEDERAL REGISTER DATE.	TYPE OF NOTICE (PROPOSED OR FINAL RULE)	REQUIREMENTS / SUBJECT	EFFECTIVE DATE
NHTSA	217	6/3/1976	Final Rule	Amended to permit the option of either a rear door, or a side door and a 10 in by 48 in rear window as school bus emergency exits (the restriction to non-engine buses was deleted). The proposed 1976 revisions were also adopted to revise emergency exit labeling on non-school buses, as well as school buses (the latter requiring more specific letter provisions).	October 27, 1976
NHTSA		6/17/1976	Correction	A small correction notice to correct heading.	
NHTSA	571.3	8/26/1976	Delay	Effective date for new school bus definition delayed from October 27, 1976	April 1, 1977
NHTSA	220	8/26/1976	Final rule	The requirement to require emergency exit operability during the force application to the roof was deleted.	April 1, 1977
NHTSA	217	2/8/1979	Final Rule	Amended on an interim final rule basis to decrease the size of the rear emergency exit for vehicles with gross vehicle weight of less than 10,000 lbs. (typically vans)	February 8, 1979
NHTSA	217	2/18/1982	Final rule	Incorporated numerous industry standards, e.g., ASTM, etc., by reference in Part 571.5	March 22, 1982
NHTSA	217	8/26/1982	Correction	Amended to correct a phrase to school bus GVWR of 10,000 lbs, "or less	August 26, 1982
NHTSA	217	11/4/1988	ANPRM	Consider amendments to increase the number, size, etc. of school bus emergency exits	NA

U.S. Bus Emergency Egress Regulation History (8)

AGENCY	FMVSS / FMCSS	FEDERAL REGISTER DATE	TYPE OF NOTICE (PROPOSED OR FINAL RULE)	REQUIREMENTS / SUBJECT	EFFECTIVE DATE
NHTSA	217	3/15/1991	Proposed Rule	Proposed to increase the number of school bus emergency exits based on capacity, instead of requiring the same number of exits for all types of buses. Options presented use different equations to calculate the number of necessary emergency exits based on the number of seating positions and the configuration of existing exits. Option A would provide that all additional exits are door exits, while Option B would require that additional exits be one or more roof hatches. Pop-out windshields were also considered, and increasing door exit and roof exit conspicuity by using retroreflective marking options were also proposed	NA
NHTSA	217	11/2/1992	Final Rule	Amended to improve school bus emergency exit provisions by considering seating capacity, and adding doors, roof hatches, and a combination, in that order. Adopted revised interior emergency exit marking requirements for buses and school buses. Adopted exterior retroreflective material marking requirements for all school bus emergency exits (ASTM Type III).	May 2, 1994
NHTSA	217	11/2/1992	Proposed Rule	Proposed to permit non-school buses to meet upgraded school bus requirements and delete the provision to provide a door for each three passengers.	NA
NHTSA	217	12/2/1992	Corrected	Corrected errors in November 1992 final rule concerning metric conversion for opening size area	NA
NHTSA	217	12/1/1993	Proposed rule	Proposed to allow emergency exit windows other than push-out type. Manufacturers would also be allowed to install vertically sliding emergency windows or an emergency exit door on school buses to satisfy the revised FMVSS 217. Mixture of windows not allowed.	NA

U.S. Bus Emergency Egress Regulation History (9)

AGENCY	FMVSS / FMCSS	FEDERAL REGISTER DATE, etc.	TYPE OF NOTICE (PROPOSED OR FINAL RULE)	REQUIREMENTS / SUBJECT	EFFECTIVE DATE
NHTSA	217	5/4/1994	Final Rule	Delayed the effective date of one section of the upgraded school bus emergency exit regulations contained in 1992 Notice, based on the interpretation of "daylight opening" due to misunderstanding of what the term meant	September 1, 1994
NHTSA	217	5/9/1995	Final Rule	Revised Title to "Bus Emergency Exits and Window Retention and Release. Amended to permit the installation of 2 emergency egress windows as an alternative to a single emergency exit door on school buses, as already required by FMVSS 217, permit non-school buses to meet certain school bus emergency exit requirements. Permitted mixture of sliding and push out windows for buses other than school buses. Ccorrected school buse marking retroreflective tape size.	May 9, 1996
NHTSA	217	3/5/1999	Proposed Rule	Proposed to amend by regulating location of wheelchair anchorages so they cannot be installed in locations where they block any emergency exit needed for school bus evacuation. Nothing would require that wheelchair positions be provided. In addition labeling requirements to prevent blocking of emergency exit openings were proposed.	NA
NHTSA	217	4/19/2002	Final Rule	Amended to reduce the likelihood that school bus wheelchair anchorages would be installed in locations that would block emergency egress. In addition, added a new requirement that doors and exits currently labeled as Emergency Doors or Emergency Exits be also labeled with "Do Not Block" in a color that contrasts with the background of the label.	April 21, 2003

U.S. Bus Emergency Egress Regulation History (10)

AGENCY	FMVSS / FMCSS	FEDERAL REGISTER DATE, etc.	TYPE OF NOTICE (PROPOSED OR FINAL RULE)	REQUIREMENTS / SUBJECT	EFFECTIVE DATE
NHTSA	217	4/22/2003	Delay in rule	Revised effective date from from April 21, 2004, and amended to respond to petitions for reconsideration concerning the volume around the side and rear emergency exit doors where wheelchairs anchorages may be located.	April 21, 2004
NHTSA	217	3/12/2004	Delay in rule	Revised effective date from April 1, 2004.	April 21, 2006
NHTSA	217	8/12/2005	Final Rule	Amended to further respond to petitions for reconsiderations relating to wheelchair anchorages where they could block emergency exits. NHTSA agreed with the petition concerning the placement of the parallelepiped in the tangent to the opening of the rear door and revised the final rule to require that anchorages that are raised, flush, or recessed in the school bus beneath the recessed beneath the parallelepiped are not permitted. The petition to allow anchorages blocking access to emergency window exits was denied. In addition, the original required "DO NOT BLOCK" label was maintained, with a phrase added to clarify that the label was required only for wheelchair anchorages	April 24, 2006
FMCSA	393	8/15/2005	Final Rule	Amend 393.62 to make the bus exit requirements compatible with the NHTSA regulations. Deleted 393.92 (emergency door marking requirements) without mention of why in Notice. NOTE: no revisions had been made to this section after June 10, 1972.	September 14, 2005

A-10

APPENDIX B. FMVSS 217 Bus Egress Requirements

COMPONENT	TEST REQUIREMENT SUMMARY	OTHER THAN SCHOOL BUSES	SCHOOL BUSES
5.1 WINDOW RETENTION	Prevent formation of a opening under specified test conditions	Yes	Yes
5.2 EMERGENCY EXITS			
Minimum Area	Unobstructed opening Area - Based on capacity	Area in cm^2 = the number of seats multiplied by 432	Not specified.
Minimum number and location	Based on capacity: # of seats		
Exit door	Number	1 rear, waived if roof hatch exit provided	At least 1. 2, or more for larger buses
Exit Window	Number	Enough to satisfy minimum area requirement with 40 percent distributed on each side and with no more than 3458 cm^2 credit for any one window; must have same number on both sides	At least 2, for all but the smaller buses Even umber on each side of bus
Roof hatch	Number	At least 1, if no rear door is provided	At least 1
	Location	None specified	For 2 hatches: at 1/4 and 3/4 of the bus length
5.3 EMERGENCY EXIT RELEASE			
	Other than School buses		
	Location of low-force areas	Figures 1 & 3	
	Type of motion	Rotary or straight	
	Magnitude (limit)	89 N	
	Location of high-force areas	Figures 2 & 3	
	Type of motion	Rotary or straight	
		Straight, perpendicular to undisturbed exit surface	
	Magnitude (limit)	267 N	
	School buses		
	Doors		
	Must be in high-force region		Fig 3A & 3D
	Type of motion		Upward from inside the bus
	Magnitude (limit)		178 N
	Windows		
	Location of low-force areas		Fig 1 & 3
	Type of motion		Rotary or straight
	Magnitude (limit)		89 N
	Location of high-force areas		Figures 2 & 3
	Type of motion		Straight, perpendicular to undisturbed exit surface
	Magnitude		178 N
	Roof Exits		
	Location of low-force areas		Figure 3B
	Type of motion		Rotary or straight
	Magnitude (limit)		89 N
	Location of high-force areas		Figure 3B
	Type of motion		Straight, perpendicular to undisturbed exit surface
	Magnitude (limit)		178 N

COMPONENT		TEST REQUIREMENT SUMMARY	OTHER THAN SCHOOL BUSES	SCHOOL BUSES
5.4 EMERGENCY EXIT OPENING				
		Other than school buses - all emergency exits	Must allow unobstructed passage of 50 x 33 cm ellipsoid before and after window retention test and within force levels specified in 5.3	
		School buses >10,000 lbs GVWR		
		Rear Doors		Must allow unobstructed passage of 1145 x 610 x 305 mm parallelepiped within force levels specified in 5.3
		Side Doors		Must allow clear opening 114 x 61 cm and within force levels specified in 5.3
		Windows		Must allow unobstructed passage of 50 x 33 cm ellipsoid and within force levels specified in 5.3
		Roof Exits		Must allow clear opening 41 x 41 cm and within force levels specified in 5.3
5.5 EMERGENCY EXIT IDENTIFICATION				
		Signage		
		Doors	"Emergency Door" or "Emergency Exit"	"Emergency Door" or "Emergency Exit" in letters at least 5 cm high
		Windows & roof exits	"Emergency Exit" & Concise operating instructions	"Emergency Exit" & concise operating instructions in letters at least 1 cm high
		Illumination	Sufficient for legibility to persons nearby	A color that contrast with the background
5.6 TEST CONDITIONS				
		Flat horizontal surface	Yes	Yes
		Temperature	70 - 85° F	70 - 85° F
		Windows installed closed and latched	Yes	Yes
		Seats, armrests, etc installed for normal use	Yes	Yes

APPENDIX C. NHTSA Motorcoach / School Bus and Other Vehicle Egress Design Regulations

Table C-1. General Emergency Exits – *Interior / Egress* (Non-door specific) (1)

REQUIREMENTS	NHTSA 49 CFR *	FRA 49 CFR **	FAA 14 CFR	USCG 46 CFR Subchapter K	SOLAS/IMO ***
TYPE LOCATION SURFACE SIZE TYPE OF RELEASE NOTIFICATION TO OPERATOR IF OPEN NOTE: Agencies vary (including NHTSA) on the type and number of doors and related requirements for various types of emergency exits, based on passenger capacity EXIT Marking: See Tables C-7 through C-12	See Tables C-2 through C-5	See Tables C-2 through C-5	*25.803 Emergency evacuation* Passengers must be evacuated in under 90 sec (demo with ½ exits disabled and under emergency lighting) *25.805 Emergency exits* Identifies exits by type, not location, depending on capacity) All aircraft must be provided with at least one exit on each side.) *25.809 Emergency exit arrangement* (a) Must be movable door or hatch in external walls of fuselage, allowing clear opening to outside. (b) Openable from inside & outside & when there is no fuselage deformation. *25.813 Emergency exit access* Each exit must be accessible to passengers and located to afford effective evacuation. Exit distribution must be as uniform as practical. (Plus additional requirements) See Tables C-2 through C-8	*116.500 Means of escape* (a) At least two exits for space accessible to passengers or regularly used by crew. One must not be watertight. (b) Two exits must be widely separated, if possible at opposite ends or sides of space. (c) Means of escape may include normal exits, emergency exits, passageways, stairways, ladders, and window. (d) The number and dimensions must be sufficient for the rapid evacuation in an emergency for the number of persons served, as determined by 116.438 (n) (2). i.e., number of fixed seats, public spaces, etc. (e) Wide enough to allow easy movement of persons wearing life jackets and with no protrusions. (f) Clear opening of door or passageway at least 32 in (81 cm), but only 28 in (71 cm) if used solely by crew. (g) No dead end passageways more than 20 ft (6m) long.	*Chap II-228.5.* At least one readily accessible enclosed stairway providing continuous shelter as means of escape, at least 900 mm (36 in) wide.

* Motorcoaches may meet certain school bus egress provisions as an alternative ** Certain requirements apply based on passenger train speed *** USCG "accepted"

C-1

Table C-1. General Emergency Exits – *Interior / Egress* (Non-door specific) (2)

REQUIREMENTS	NHTSA 49 CFR *	FRA 49 CFR**	FAA 14 CFR	USCG 46 CFR Subchapter K	SOLAS/IMO ***
TYPE LOCATION SURFACE SIZE TYPE OF RELEASE NOTIFICATION TO OPERATOR IF OPEN NOTE: Agencies vary (including NHTSA) on the type and number of doors and related requirements for various types of emergency exits, based on passenger capacity EXIT Marking: See Tables C-7 through C-12				*116.500 Means of escape* (h) Maximum allowable walking distance from most remote point to nearest exit is 150 ft (46 m). (i) Capable of being opened by one person, from either side, in any lighting or in darkness.	

* Motorcoaches may meet certain school bus egress provisions as an alternative   ** Certain requirements apply based on passenger train speed   *** USCG "accepted"

Table C-2. Emergency Doors / Doors Used as Emergency Exits- *Interior Egress* (1)

REQUIREMENTS	NHTSA 49 CFR *	FRA 49 CFR**	FAA 14 CFR	USCG 46 CFR Subchapter K	SOLAS/IMO***
DOOR / SIDE DOOR / REAR DOOR Type Location Surface Size Type Of Release Notification to Operator if open NOTE: Agencies vary (including NHTSA) on the type and number of doors and related requirements for various types of emergency exits, based on passenger capacity EXIT Marking: See Tables C-7 through C-12	571.217 5.2 *Other than School bus* 5.2.2.2 1 side door per each 3 passenger seats and at least one rear exit. If no rear exit, then at least one roof hatch (see also Table C-5) 5.3. *School bus* 1 left side door and 1 rear door with a push-out window Additional exits required based on bus capacity (See Table C-1). 5.4.2.1 *School bus* (2) Opening at least 114 cm (45 in) high and 61 cm (24 in) wide. 5.2.3.1 *(School bus)* (a) Rear door opens outward and is hinged on the right side. 5.3.1 No more than 2 release mechanisms.	238.235 *Doors* (b) & 238.439 (Tier II) *Doors (a):* (b) At least two exterior side doors. Clear opening at least 30 in (76 cm) wide by 74 in (188 cm) high. Power side doors in vestibules: Manual override device that: (1) Permits opening door from inside the car without power (2) Is adjacent to door it controls. (3) Allows use of device from inside car without an implement. 238.439 *Doors (Tier II)* (c) Display status of each door to the crew in the operating cab. (d) Connect doors to emergency back-up power system. (f) Have a kick-out panel, pop-out window or similar egress means.	25.807 *Emergency exits.* (a) (1) Type I. A floor-level exit with a rectangular opening not less than 24 in wide by 48 in high, with corner radii not greater than 8 in. .... (g) No more than 45 passengers per Type I exit in each side of the fuselage. Other requirements for Type II & IV and Type A-C door exits 25.809 *Emergency exit arrangement* (a) Must be movable door or hatch in external walls of fuselage, allowing clear opening to outside. (b) Openable from inside & outside & when there is no fuselage deformation. 25.813 *Emergency exit access.* Each exit must be accessible to passengers and located to afford effective evacuation. Exit distribution must be as uniform as practical. Other requirements for Type II, IV and Type A-C door exits	116.435 *Doors* (b)(2) Cable of being opened by one person, from either side. (6) The maximum width of an individual door must not exceed 48 in (100 mm).	*See Table C-1*

* Motorcoaches may meet certain school bus egress provisions as an alternative ** Certain requirements apply based on passenger train speed *** USCG "accepted"

Table C-2. Emergency Doors / Doors Used as Emergency Exits- *Interior Egress (2)*

REQUIREMENTS	NHTSA 49 CFR *	FRA 49 CFR **	FAA 14 CFR	USCG 46 CFR Subchapter K	SOLAS/IMO***
DOOR / SIDE DOOR / REAR DOOR (con't) Type Of Release NOTE: Agencies vary (including NHTSA) on the type and number of doors and related requirements for various types of emergency exits, based on passenger capacity EXIT Marking: See Tables C-7 through C-12			*Varies depending on type*	*116.500 Means of escape* (h) Maximum allowable walking distance from most remote point to nearest exit is 150 ft (46 m). (i) Capable of being opened by one person, from either side, in any lighting or in darkness. *116.435 Doors* (b)(2) Cable of being opened by one person, from either side. (6) The maximum width of an individual door must not exceed 48 in (100 mm).	

* Motorcoaches may meet certain school bus egress provisions as an alternative ** Certain requirements apply based on passenger train speed *** USCG "accepted"

Table C-3. Emergency Window Exits – *Interior Egress*

REQUIREMENTS	NHTSA 49 CFR *	FRA 49 CFR **	FAA 14 CFR	USCG 46 CFR Subchapter K	SOLAS/IMO***
NUMBER LOCATION SIZE RELEASE TYPE NOTE: Agencies vary (including NHTSA) on the type and number of emergency exit windows and related requirements for various types of emergency exits, based on passenger capacity EXIT Marking: See Figures C-6 through C-12	*571.217* *5.2.2.1 (Bus)* Unobstructed opening surface area in total square cm. at least 432 x number of seating positions. At least 40 % of required surface area must on each side. No 1 emergency exit to be more than 3,458 cm² (536 in) of the total area. Sliding and push-out emergency exit windows are permitted. *5.3.2 Buses* Open manually to required size by a single occupant using conforming high and low force applications. *5.4.1 (Bus)* (b) Opening of rotating ellipsoid with major axis of 20 in (51 cm) and minor axis of 13 in (33 cm). *5.2.3.2 (School bus)* (c) Emergency exit windows. Even number of windows equally distributed on each side. No horizontally sliding emergency windows. Both sliding and push-out exit windows not allowed, except bus with a push-up window in a rear door and side sliding windows . *5.2.3.1 (School bus)* (b) Unobstructed opening of rotating ellipsoid with major axis of 20 in (51 cm) and minor axis of 13 in (33 cm). (b) Open manually to required size by a single occupant using conforming high and low force application. No remote control permitted Exit interlock required	*238.113 Emergency Window Exits* (a) (1) At least four emergency window exits on each car level, staggered or one exit on each end of each side of car. 3)(a) Permit rapid, easy removal of window from inside car without special tool or an implement. *238.113* (new equipment) (b) Unobstructed opening at least 26 in (66 cm) wide by 24 in (61 cm) high (new equipment)	NONE	*116.500* *Means of escape* (c) The two means of escape required in 116.500 (a) & (b), listed in Table 1, can include windows. *116.500* *Means of escape* (e) No protrusions and wide enough for easy movement of persons wearing life jackets. (f) Clear opening of at least 32", but only 28" if used solely by crew.	NONE

* Motorcoaches may meet certain school bus egress provisions as an alternative   ** Certain requirements apply based on passenger train speed   *** USCG "accepted"

Table C-4. Doors and Windows – Rescue Access - *Exterior*

REQUIREMENTS	NHTSA 49 CFR *	FRA 49 CFR **	FAA 14 CFR	USCG 46 FR Subchapter K	SOLAS/IMO***
DOORS Number Location Size Release Type EXIT Marking: See Figures C-6 through C-12	*571.217* *Motorcoaches – NONE* *5.4.2.1 School bus* (2) Opening at least 114 cm (45 in) high and 61 cm (24 in) wide. *5.3.3 School bus* Openable from the outside using specified high and low force limits.	*238.235 Doors* (b) At least two exterior side doors. Clear opening at least 30 in (76) wide by 74 in (188 in) high. Power side doors in vestibules: Manual override device that: (1) Permits opening door from inside the car without power. (2) Is adjacent to door it controls. (3) Allows use of device from outside car without an implement.	*NONE*	*NONE*	*NONE*
WINDOWS Number Location Size Release Type EXIT Marking: See Figures C-6 through C-12	*571.217* *Motorcoaches – NONE* *5.4.2.1 School bus* (2) Opening at least cm (45 in) high and 61 cm (24 in) wide. *5.3.3 School bus* Openable from the outside using specified high and low force limits. * Motorcoaches may meet certain school bus egress provisions as an alternative.	*238.114 Rescue Access Windows* (a) (1) At least two access window exits on each main car level, towards mid point of car to the extent practicable (2) Unobstructed opening at least 2 per car side at midpoints (3) (c) At least 26 in (66cm) wide by 24 in (61 cm) high. *(new equipment)* *(e)* Permit rapid, easy removal of window	*NONE*	*NONE*	*NONE*

* Motorcoaches may meet certain school bus egress provisions as an alternative   ** Certain requirements apply based on passenger train speed   *** USCG "accepted"

Table C-5. **Emergency Roof Exit / Access Roof** –*Interior and Exterior*

REQUIREMENTS	NHTSA 49 CFR *	FRA 49 CFR **	FAA 14 CFR	USCG 46 FR Subchapter K	SOLAS/IMO***
INTERIOR Type Number Location Size Release Type	*571.217* *S.5.2.2.2 (Bus)* Roof hatch hinged on forward side, provided in the rear half of bus, if no rear door provided. Must meet requirements of S.5.3-5.5 when bus is overturned *5.2.3 (School bus)* Type & number chosen per manufacture option per Capacity Tables (see 5.2.3.1-5.2.3.2). *5.4.2 (School bus)* (b) Opening 16 by 16 in (41 by 41 cm) *5.2.3 (School bus)* Openable from inside *S 5.3 (All)* Not more than 2 release mechanisms are required to open the exit		*NONE*	*NONE*	*NONE*
EXTERIOR Type Number Location Size Release Type	*Exterior operation not required for motorcoaches.* S 5.2.3.2 *School buses* (b) Exterior operation is required (of the interior hatch meeting the above requirements	*238.123 Roof Access* - hatch or soft spot (new equipment) (a) At least two (Unobstructed from the inside access points on each car), one in each half of the car, staggered *238.429(New and Existing)* At least one hatch or soft spot 18 in (46 cm) by 24 in (61 cm) *238.123 New Equipment* Opening at least 26 in (66 cm) wide by 24 in (61 cm) high *238.113 & 238.429* *Both 238.1243 and 238.429:* Openable from inside by hatch or soft spot that can be cut from outside			

* Motorcoaches may meet certain school bus egress provisions as an alternative   ** Certain requirements apply based on passenger train speed   *** USCG "accepted"

Table C-6. Interior (Egress) Exit Marking / Signs & Instructions – Emergency Doors / Doors Used as Emergency Exits-

REQUIREMENTS	NHTSA 49 CFR*	FRA 49 CFR ** (238.235 refers back to 239.107)	FAA 14 CFR	USCG 46 CFR Subchapter K	SOLAS/IMO***
INTERIOR Type Location Minimum Criteria	571.217 5.5.3 School Bus (a) "Emergency Exit" in letters at least 5 cm (2 in) high of a color that contrasts with background, at top of, directly above, or at bottom of door. (b) Instructions on how to unlatch and open door. Located within 15 cm (6 in) of release, in letters at least 1 cm (0.5 in) high and of color that contrasts with background. (d) Label on inside directly beneath or above each door, in letters at least 2.5 cm (1 in) high "DO NOT BLOCK", in color that contrasts with label background. 5.5.2.1 (Bus) Each marking legible in normal nighttime illumination of bus interior, to occupants having corrected visual acuity of 20/40.	239.101 Em Prep Plan (a) written plan with following elements… (7) Passenger Safety Program (ii) Passenger Awareness Program Activities Each RR conspicuously and legibly post emergency instructions inside all passenger cars (e.g., on car bulkhead signs, seatback decals, or seat cards) 239.107 Exits (a) (1) all door exits intended for emergency egress are either lighted or conspicuously and legibly marked with luminescent material on the inside of the car and that clear and understandable instructions are posted at or near such exits.	25.811 Emergency exit marking. (a) Conspicuously mark each emergency exit, its means of access, and its means of opening. (b) Identity and location of each emergency exit recognizable from a distance equal to width of cabin. (c) Provide means to assist occupants in locating exits in dense smoke. (d) Mark each emergency exit with sign visible when approaching in the main passenger aisle (or aisles). There must be— (1) A passenger emergency exit locator sign above aisle near each emergency exit. (2) Emergency exit sign next to each emergency exit. (3) A sign on each bulkhead or divider that prevents fore and aft vision along passenger cabin to indicate emergency exits beyond and obscured by the bulkhead or divider. (e) Location of operating handle and instructions for opening exits from inside: (1) Each exit, a marking readable from (76 cm) 30 in. conspicuously marked per different handle types. (g) Each sign required by paragraph (d) of this section may use the word "exit" instead of "emergency exit".	116.500 Means of escape (j) A means of escape not readily apparent from inside the space must be adequately marked in accordance with Sec. 122.606. 116.520 Emergency evacuation plan. (a) Possible casualties involving fires or flooding. (b) Procedures for evacuating all affected. (1) Accessible refuge for capacity of vessel. (2) Two escape means 122.606 Escape hatches & emergency exits. Must be marked on both sides in clearly legible letters at least 2 in high: "EMERGENCY" 122.610. Watertight doors and hatches. Must be marked in clearly legible letters at least 1in (2.5 cm) high	Chap II-2 28.5.10. All escape route signs photoluminescent material or marked by lighting. 28-1.1.4/5 Escape route shall be unobstructed, and accessible from all public spaces. Floor coverings shall be secured.

* Motorcoaches may meet certain school bus egress provisions as an alternative   ** Certain requirements apply based on passenger train speed   *** USCG "accepted"

C-8

Table C-7. Exterior (*Rescue Access*) Marking / Signs & Instructions - Emergency Doors / Doors Used as Emergency Exits

REQUIREMENTS	NHTSA 49 CFR*	FRA 49 CFR** (238.235 refers back to 39.107	FAA 14 CFR	USCG 46 CFR Subchapter K	SOLAS/IMO ***
TYPE LOCATION MINIMUM CRITERIA INSTRUCTIONS	571.217 5.5.3 School Bus (a) "Emergency Exit" in letters at least 2 in high of a color that contrasts with background, at top of or directly above door. (c) Each opening outlined around perimeter with red, white, or yellow retroreflective tape at least 1 in wide and meet conditions in 49 CFR 571.131.S6.	238.114 Marking and instructions Each (exterior) rescue access window marked w/ retroreflective material. A unique and easily recognizable symbol, sign, or other conspicuous marking to identify each such window. Legible and understandable window access instructions for removing such window at or near each such window 239.107 Exits (a) Marking (2) That all door exits intended for emergency access by emergency responders are marked with retroreflective material. Clear understandable instructions are posted at each such door.	25.811 *Emergency exit marking*. (f) Each emergency exit required to be openable from outside, and its means of opening, must be marked on outside of airplane.	116.500 *Means of escape* (i) A means of escape not readily apparent from outside the space be adequately marked per 122.606. 122.606 *Escape hatches & emergency exits*. Must be marked on both sides in clearly legible letters at least 5 cm (2 in) high: "EMERGENCY EXIT" Sec. 122.610 *Watertight doors and hatches*. Must be marked in clearly legible letters at least 2.5 cm (1in) high:	NONE

* Motorcoaches may meet certain school bus egress provisions as an alternative   ** Certain requirements apply based on passenger train speed   *** USCG "accepted"

Table C-8. Interior (*Egress*) Exit Marking Signs and Instructions - Emergency Window Exits

REQUIREMENTS	NHTSA 49 CFR*	FRA 49 CFR**	FAA 14 CFR	USCG 46 CFR Subchapter K	SOLAS/IMO***
INTERIOR SIGNS (*Egress*) Type Location Minimum Criteria Instructions	571.217 5.5.1 Each marking legible in normal nighttime illumination of bus interior, to occupants having corrected visual acuity of 20/40 at three specific locations. Instructions on how to unlatch and open window, e.g., "Lift to Unlatch, Push to Open;" or "Turn Handle, Push Out to Open. Located within 16 cm (6 in) of mechanism 5.5.3 *School Bus* (a) "Emergency Exit" in letters at least 5 cm (2 in) high of a color that contrasts with background, at top of, directly above, or at bottom of window. (b) Instructions on how to unlatch and open window, e.g., "Lift to Unlatch, Push to Open;" or "Turn Handle, Push Out to Open. Located within 15 cm (6 in) of release, in letters at least 1 cm (0.5 in) high of color that contrasts with background. (d) Label on inside directly beneath or above each window, in letters at least 2.5 cm (1 in) high "DO NOT BLOCK" in color that contrasts with label background.	238.113 *Requirements for new or rebuilt equipment* (d) Marking. (1) Each emergency window conspicuously and legibly marked w/ luminescent material & post clear and legible operating instructions at or near each such exit. 239.101 *Em Prep Plan* (a) Written plan w/ following elements (7) Passenger Safety Program (ii) Passenger Awareness Program Activities Each RR: conspicuously and legibly post emergency instructions inside all passenger cars (e.g., on car bulkhead signs, seatback decals, or seat cards)	*Same as for Doors*	116.500 *Means of escape* (j) An escape widow not readily apparent from inside the space must be adequately marked in accordance with Sec. 122.606. 122.606 *Escape hatches & emergency exits*. Escape windows must be marked on both sides in clearly legible letters at least 5 cm (2 in) high: "EMERGENCY EXIT"	NONE

* Motorcoaches may meet certain school bus egress provisions as an alternative   ** Certain requirements apply based on passenger train speed   *** USCG "accepted"

Table C-9. Exterior (*Rescue Access*) Signs & Instructions – Rescue Access Windows

REQUIREMENTS TYPE LOCATION MINIMUM CRITERIA INSTRUCTIONS.	NHTSA 49 CFR *	FRA 49 CFR **	FAA 14 CFR	USCG 46 CFR Subchapter K	SOLAS/IMO***
	571.217 5.5.3 *School Bus* (a) "Emergency Exit" in letters at least 5 cm (2 in) high of a color that contrasts with background, at top of, directly above, or at bottom. (b) Instructions on how to unlatch and open window, e.g., "Lift to Unlatch, Push to Open;" or "Turn Handle, Push Out to Open. Located within 15 in (6) in of release, in letters at least (1 cm) 0.5 in high and of color that contrasts with background. (c) Each opening outlined around outside perimeter with red, white, or yellow retroreflective tape at least 2.5 cm (1 in) wide which meets conditions in 49 CFR 571.131.S6 (ASTM "Type III)	238.114 *Requirements for new or rebuilt equipment* (d) Marking (2) Each window intended for emergency access by emergency responders for extrication of passengers marked with a retroreflective, unique, and easily recognizable symbol or other clear marking (ASTM "Type I"). Clear & understandable window-access instructions posted either at each such window or at each end of the car.	NONE	116.500 *Means of escape* (j) An escape widow not readily apparent from outside the space must be adequately marked in accordance with Sec. 122.606. 122.606 *Escape hatches & emergency exits.* Escape windows must be marked on both sides in clearly legible letters at least 2 in high: EMERGENCY EXIT	NONE

* Motorcoaches may meet certain school bus egress provisions as an alternative   ** Certain requirements apply based on passenger train speed   *** USCG "accepted"

Table C-10. Interior and Exterior Roof Exit / Access – Marking, Signs and Instructions

REQUIREMENTS	NHTSA 49 CFR	FRA 49 CFR	FAA 14 CFR	USCG 46 CFR Subchapter K	SOLAS/IMO***
INTERIOR Type Location Minimum Criteria	5.5.3. *School Bus* (a) "Emergency Exit" in letters at least 2 in high located on the inside surface or within 30 cm of the roof exit and of color that contrasts with background... (b) Instructions on how to unlatch and open exit, e.g., "Lift to Unlatch, Push to Open;" or "Turn Handle, Push Out to Open."	NONE	NONE	NONE	NONE
EXTERIOR Type Location Minimum Criteria	5.5.3. *School Bus* (a) Doors are marked with letters at least 2 in high and of color that contrasts with background. (c) Each opening outlined around outside perimeter with red, white, or yellow retroreflective tape at least 2.5 cm (1 in) wide which meets conditions in 49 CFR 571.131.S6 (ASTM "Type III")	238.123 *Requirements for new or rebuilt equipment* (d) Marking. (1) Each roof access point marked outline of 2.5 cm (1 in) of conspicuous retroreflective material (ASTM "Type I"). Understandable and legible operating instructions at or near each such exit.	NONE	NONE	NONE

* Motorcoaches may meet certain school bus egress provisions as an alternative ** Certain requirements apply based on passenger train speed *** USCG "accepted"

Table C-11. Emergency Lighting

REQUIREMENTS	NHTSA 49 CFR *	FRA - 49 CFR **	FAA 14 CFR	USCG 46 CFR Subchapter K	SOLAS / IMO***
LOCATION MINIMUM CRITERIA Illuminance Back-up power Angle of operation Shock Operation time period	*No requirements for either motorcoaches or school buses* *Exit sings must be visible only under normal illumination*	238.115 Emergency Lighting (NEW equipment only) a) Each level of a multi-level passenger car. b) (1) Minimum, average illuminance = 1 fc. at floor level, adjacent to each exterior door and each interior door providing access to an exterior door (such as a door opening into a vestibule); 2) Minimum, average illuminance = 1 fc. 25 in above floor level along the center of each aisle & passageway; (3) Minimum illuminance = 0.1 fc measured 25 in above floor level at any point along the center of each aisle and passageway; and (4) Back-up power system capable of: (i) Operating in all equipment orientations within 45° of vertical; (ii) Operating after initial shock of collision or derailment individually applied accelerations: (A) Longitudinal: 8g; (B) Lateral: 4g; and (C) Vertical: 4g; and (iii) Operating all emergency lighting for at least 90 minutes without a loss of more than 40% of minimum illuminance specified in (b) 239.107 Exits (a) (1) Permits "lighted" as marking at doors, as alternative to luminescent marking However, no criteria are provided (issued prior to 238.115).	25.812 Emergency lighting. (a) emergency lighting system. independent of main lighting system: (1) illuminated emergency exit marking and locating signs. (i) general cabin illumination (ii) interior lighting in emergency exit areas. (iii) floor proximity escape path marking. (2) Exterior lighting. (b) Emergency exit signs (1) For 10 or more passenger seats. (i) Each emergency exit locator sign and each emergency exit marking sign: - Red letters at least 1 ½ in high on illuminated white background at least 21 in square excluding letters. - Lighted background-to-letter contrast at least 10:1. - The letter height to stroke-width ratio not more than 7:1 or less than 6:1. These signs must be internally electrically illuminated with a background brightness of at least 25 footlamberts and high-to-low background contrast no greater than 3:1. (ii) For each passenger emergency exit sign required by §25.811 (d)(3). same as (b) (1) Signs internally electrically illuminated or self-illuminated by other than electrical, and initial brightness of at least 400 microlamberts.	120.432 Emergency lighting. (a) Adequate emergency lighting fitted along line of escape to main deck from all passenger and crew spaces below main deck. (b) Must automatically actuate upon failure of the main lighting system. If no single source of power for emergency lighting, must have individual battery powered lights which are: (1) Automatically actuated upon loss of normal power; (2) Not readily portable; (3) Connected to an automatic battery charger; (4) Of sufficient capacity to provider at least 2 hours of continuous operation.	*Chap II-1 42 Emergency Electrical Power.* 1.1. Self-contained emergency source of power 2.1 Emergency lighting supply all services to essential for safety simultaneously for 2.2 36 hrs certain locations including alley ways, stairways and exits

* Motorcoaches may meet certain school bus egress provisions as an alternative ** Certain requirements apply based on passenger train speed *** USCG "accepted"

C-13

Table C-12. Exit Path Marking (Low Location / Low Level)

REQUIREMENTS	NHTSA 49 CFR *	FRA 49 CFR **	FAA 14 CFR	USCG 46 CFR Subchapter K	SOLAS/IMO***
FLOOR PROXIMITY / LOW LEVEL Type Location Minimum Criteria Illuminance / Luminance Back-up power Operation time period	*NONE*	*NONE*	25.81 *Emergency lighting.* (d) The floor of the passageway leading to each floor-level passenger emergency exit, must be provided with illumination that is not less than 0.02 foot-candle measured along a line that is within 6 inches of and parallel to the floor and is centered on the passenger evacuation path. (e) *Floor proximity emergency escape path marking* must provide emergency evacuation guidance for passengers when all sources of illumination more than 4 feet above the cabin aisle floor are totally obscured. In the dark of the night, the floor proximity emergency escape path marking must enable each passenger	*NONE* *(But U.S ships traveling in international waters comply with SOLAS/ IMO*	Chap II-22 8.5.10 Escape path must be marked lighting or PL strip indicators not more than 0.3 m (12in) above deck *IMO Resolution A752 (18) Annex* 4.5 IMO symbols be in all low-level lighting (LLL) leading to muster stations. 6.1 All escape route signs be photo luminescent (PL) materials or marked by lighting fitted in lower 300 mm (12 in) of bulkhead. 6.2 LLL exit signs at all exits, within lower 300 mm (12 in) on side of door exits where handle is located. 6.3 All sign colors should contrast with background. 7.1 All PL strips be no more than 75 mm (3 in) wide. If less, luminance must be increased proportionally to compensate. 7.2 At least 15 mcd/m² measured 10 min after removal of external illuminating sources & continue to provide at least 2 mcd/m² for 60 min

* Motorcoaches may meet certain school bus egress provisions as an alternative   ** Certain requirements apply based on passenger train speed   *** USCG "accepted"

APPENDIX D. FMCSA Bus / Motorcoach Safety Brochure (1)

Lift Window Instructions

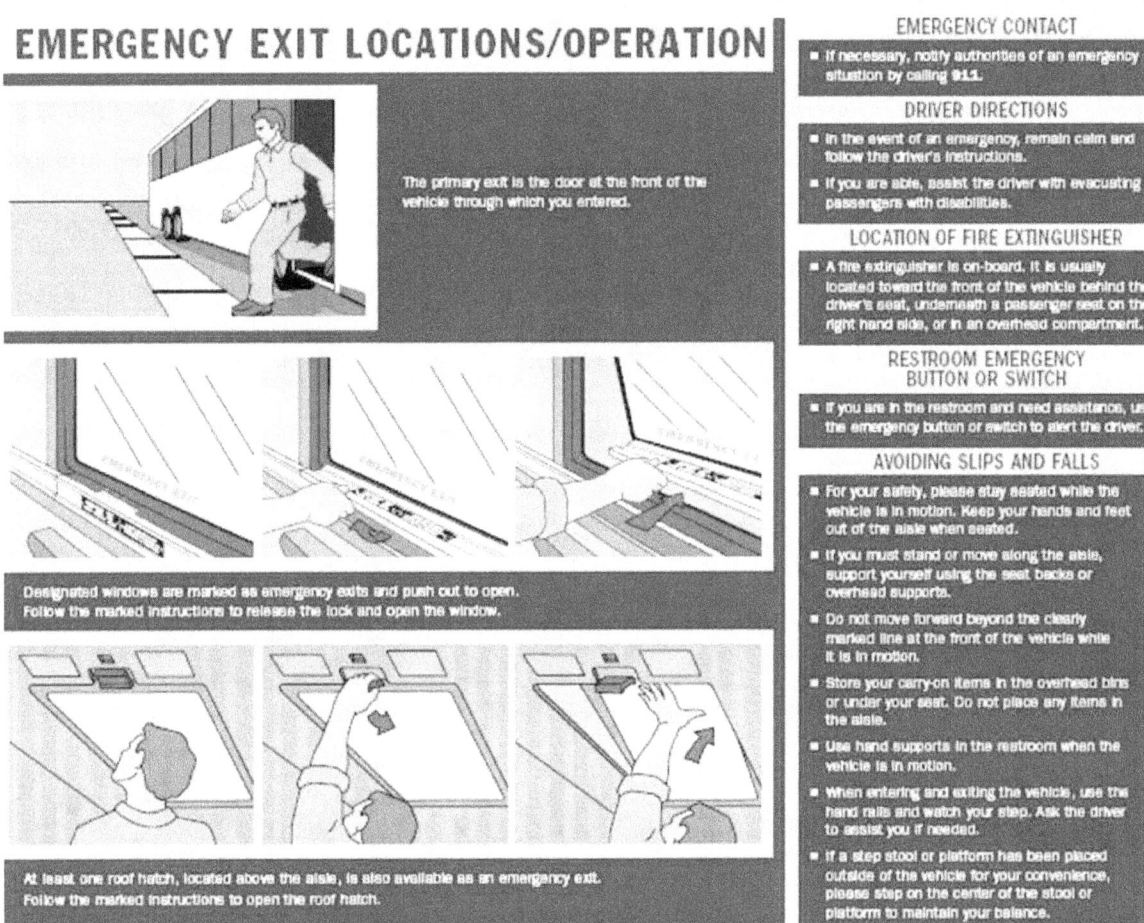

Bus motorcoach safety brochure (2)

Pull Window Instructions

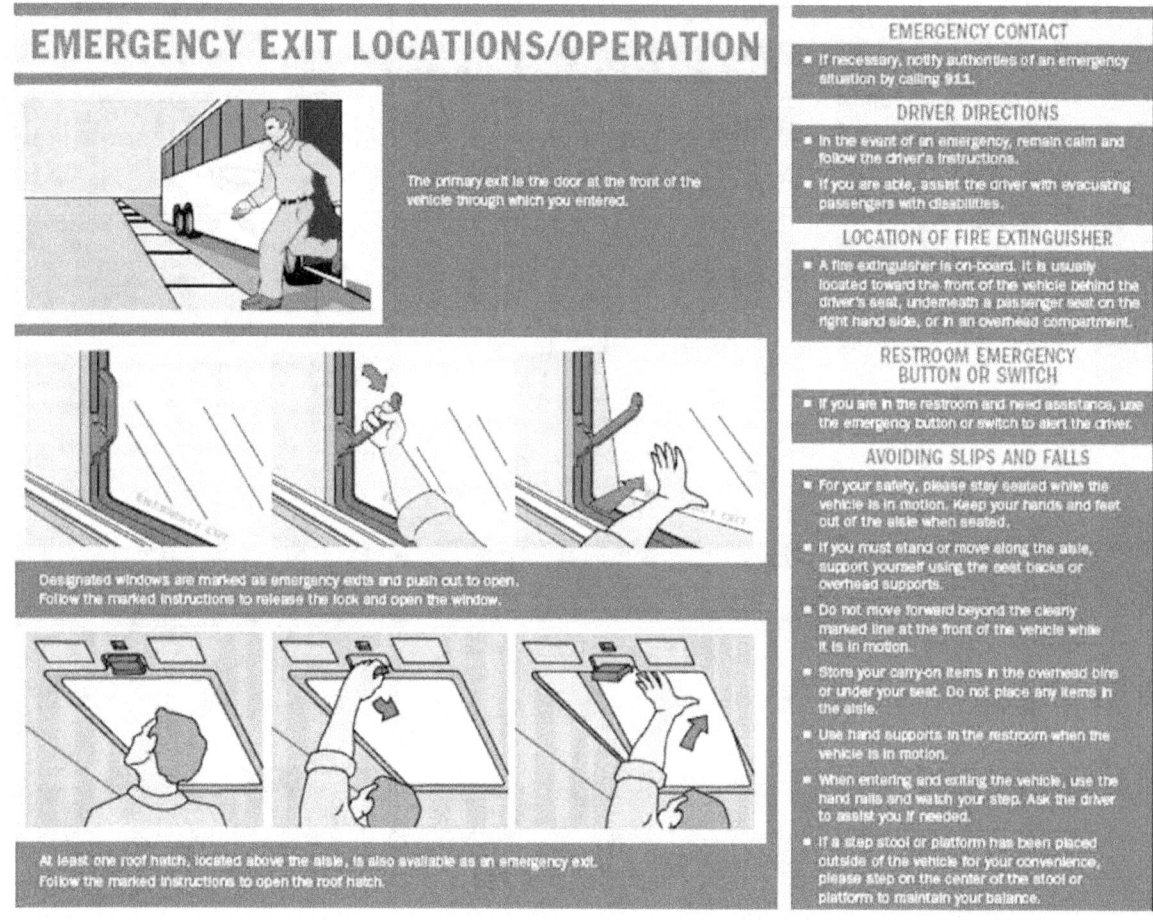

www.ingramcontent.com/pod-product-compliance
Lightning Source LLC
Chambersburg PA
CBHW080243180526

45167CB00006B/2394